Creative Machines

Creative Machines

AI, Art & Us

Maya Ackerman

Published by John Wiley & Sons, Inc., Hoboken, New Jersey.
Published simultaneously in Canada.

For general information on our other products and services or for technical support, please contact our Customer Care Department within the United States at (800) 762-2974, outside the United States at (317) 572-3993 or fax (317) 572-4002.

Wiley also publishes its books in a variety of electronic formats. Some content that appears in print may not be available in electronic formats. For more information about Wiley products, visit our web site at www.wiley.com.

Library of Congress Cataloging-in-Publication Data applied for:

Hardback ISBN: 9781394316267
ePDF ISBN: 9781394316281
epub ISBN: 9781394316274
oBook ISBN: 9781394321506

Author Photo: © DolinPhoto.com
Cover Design: Wiley

Printed and bound by CPI Group (UK) Ltd, Croydon CR0 4YY
C9781394316267_280825

To my father, Efim, who always believed in me – you are my rock.

To my husband, David, and my son, Alexander – you are my heart. Everything is for you.

Contents

Introduction

Born of sparks, they reignite
Machines unleashed to boundless flight
Weaved of echoes from the past
What horizons call to us?

At the end of 2022, I'd wake up every morning with the uneasy feeling that I had hallucinated my entire life. For nearly a decade, my friends and I had been quietly building creative machines. Then, almost overnight, our corner of the world turned into the main stage. The same playground where we'd once worked in obscurity was suddenly overrun with venture capitalists, journalists, influencers, and newly declared prophets of the AI future.

This text is not the most comprehensive overview of generative AI. It's not an attempt to download everything that can be said about this marvelous space. Instead, it is an insider's perspective from someone who had the cosmic fortune to engage in the making of creative machines nearly a decade before their proliferation.

I've always lived between worlds. I've sung Italian arias in vast, echoing spaces. Sat down at random pianos in public and let my hands play whatever came. Had my music played on the big screen. Painted, composed, written poetry.

And yet, somehow, life plunged me into computer science.

At seventeen, I stepped into a world of logic and precision, where sound became frequency and color became data. I spent a decade immersed in it – studying math, coding, machine learning. It was structured. It was orderly. It was exactly what I needed at the time.

But the desire for creative expression never left me.

On a warm day on a San Diego beach in my early thirties, I met Harold Cohen. The man who built AARON – one of the first creative machines. In that moment, something clicked. It felt like I had been looking for creative machines my entire life. A perfect marriage of my favorite things. A world fresh with possibilities.

Without hesitation, I reshaped my life around them. I left behind the world of theoretical machine learning and stepped fully into the strange, electric space where art and artificial intelligence meet – not just as a researcher, but also as an entrepreneur, launching one of the earliest generative AI startups.

These systems have also become part of my own creative process. I use *Midjourney* to generate sketches that inspire my art. I use *LyricStudio* – a tool I built in my startup – to shape my lyrics. The poems scattered across this book were made in collaboration with creative machines.

But no matter how much I work with AI, the real world always calls me back.

Back to the resonance of my piano.

Back to the texture of paint under my fingernails.

Back to the scent of flowers as I collect materials for my art.

Because at the core of it all, this world – the physical, sensory, *human* world – is what life is about.

And now, all of us on this blue-green planet stand at the edge of something new. Creative machines have broken free from academia's ivory towers. Now, they roam among us – composing, painting, imagining – creating.

This is not just another wave of automation. It's a pivotal moment, and we have choices to make. We cannot approach this moment with old thinking. It cannot be only about efficiency and profit. The possibilities before us are far greater.

If we get this right, these AI systems will lead us to a fuller human life, a more vibrant existence. We have before us the opportunity to reach further into the creative realm than ever before.

But we have to go there on purpose. We cannot afford blind optimism, assuming it will all work out. We cannot afford passive acceptance, believing that nothing can be done.

That is what I hope to offer in these pages: not just an understanding of creative machines, but a way to see them more clearly. A way to step into this future with intention.

The waves of change are already here. And this book is a vessel – a way to move through uncertainty, expand our options beyond the status quo, and decide what kind of future we want to build.

Because, in the end, the age of creative machines isn't just about technology – it's about us.

What Are Creative Machines?

What are creative machines? Where did they come from, and what hopes and dreams fueled their inventors? In this first part of the book, we'll dive into the messy, thrilling questions at the heart of AI's creative mind.

Our gateway into this world is AARON – one of the first AI systems to create art. Through the vision of its creator, Harold Cohen, we'll explore the complex bond between humans and the creative machines we dare to build.

Next, we'll tackle some foundational questions: What is creativity? How does it emerge in humans? And how do hallucinations – those strange leaps of the mind – bind the creative powers of both humans and machines? Finally, we'll dive into the origins of these machines in the towers of academia, and the great AI awakening that took place in late 2022. Come along as we peel back the layers of creativity and explore the roles of machines in this unfolding story.

AARON the Machine Painter

The Only Voice of Reason

The only voice of reason,
Unheard, unseen.
I am the creative breath,
Brushing through the wind.

It wasn't OpenAI or Google Brain that first discovered the marvelous terrain of creative machines. Even the most triumphant industry players joined the parade only long after the pioneers had already blazed the trail. By the time that Microsoft, Google, and IBM arrived on the scene, the quiet explorers had long since proved that the land of machine creativity was fertile.

The true pioneers of machine creativity worked humbly, far from the spotlight. Driven by nothing but passion and the pursuit of knowledge, these curious minds ventured into uncharted territories. Like all true explorers, they pressed forward without knowing if their efforts would ever bear fruit.

These early adventurers weren't overconfident warriors, guns blazing, arrogantly trampling over creative traditions. Instead, many were artists – respected figures in their fields – risking their reputations to carve out this new domain.

I first encountered one of these passionate pioneers back in 2015. My story begins on a warm San Diego beach. The sun glistened across the ocean, framed by palm trees and the purple flowers that define the paradise of this splendid coast. I'd been reluctantly attending the workshop on Information Theory and Applications all week, and it was the same routine of machine learning techniques and mathematical proofs.

After over a decade of education and preparation, I had just landed a job as a professor of computer science. Such a big title! Shouldn't I have been excited? I'd worked so hard to achieve this. Like a marriage that had withered too soon, my passion for machine learning, which had once consumed me through my Master's and PhD, had all but faded.

Where I expected certainty and confidence in my new position, I felt more lost and confused than ever. Even against the landscape of the San Diego paradise, a wave of melancholy washed over me.

In my hotel room, I felt paralyzed by doubt. *What have I done all this for? Why did I devote a decade to studying something that now feels so meaningless?* Conversations failed to engage me. The talks bored me. I drifted through the conference in a fog, disconnected, lost, and burdened by my own indifference.

I was yearning for something I couldn't name. My lifelong passion for music ached in my bones, but I was at a loss for what to do with it. It was becoming painfully clear that while I loved music, I had no desire to seek a musician's career. Music needed to remain my lover, a passionate side fling full of life and fire.

But the search for fulfillment felt aimless. *What am I looking for? Does it even exist?* I was starting to believe that I might never find what I was seeking.

Then, while glancing over the workshop schedule, something jumped out at me. A small session, tucked away on the last day of the workshop, titled "Computational Creativity."

Something about art and music. I made a mental note to attend.

Walking into the dusty, dimly lit lecture hall, I took a seat near the back. Close to the door. I didn't want to be trapped if the presentations were as boring as usual.

I settled into my chair. An older man I had never heard of named Harold Cohen was leading the presentation. One by one, beautiful paintings flashed on the screen: abstract flowers in vivid purples, blues, and yellows, and more abstract shapes in bright pigments collaged against one another.

Then, something extraordinary happened.

Cohen suddenly raised his voice, shouting: "And I was the only voice of reason. Saying, *I'm* the creative one!"

I straightened in my chair and started paying attention. His passion was palpable, his emotion raw and infectious. What had upset him so deeply?

I quickly pieced it together.

Cohen had created an automated painter, AARON, which had produced these impressive works of art that I had just seen. And some people had started to refer to the program, AARON, as a creative entity. Many, in fact, argued that AARON – Cohen's own creation – was proof that machines, lifeless computer programs, could be creative.

And this made Harold very, very angry.

Tingles traveled through me, head to toe. *Creative machines!* Where had they been all my life?

The Birth of a Machine Artist

Harold Cohen didn't fit the mold of an AI pioneer. He wasn't a computer scientist. He wasn't a programmer. He was an artist – brilliant, world-renowned, and celebrated for his explosive use of color and bold compositions. In 1966, he represented Britain at the Venice Biennale, one of the most prestigious art events in the world. Cohen had arrived. He was at the top of his field.

But within two years, something shifted. In 1968, Cohen found himself on the other side of the world, at the University of California, San Diego – the very campus where, decades later, I sat in the audience, enthralled by his passion. It was there, almost by accident, that he stumbled into computer programming.

By 1971, at Stanford University's Artificial Intelligence Laboratory, Cohen began building AARON, a machine that could make art. But not just any art – art that could stand on its own, that could hang in galleries. This project wasn't just about machines; it was about Cohen himself. In an audacious act of self-exploration, he set out to crack open his own mind and decode the mystery of his creative process.

Capturing the depth of Harold Cohen's creativity in lines of code was no simple task. It became a decades-long obsession. AARON's earliest works in the 1970s were humble – simple, monochrome line drawings that felt like a child's first tentative sketches. But by the 1980s, AARON had matured. It began drawing intricate forms – human figures, plants, everyday objects – capturing the natural world with surprising sensitivity.

For years, AARON acted as Cohen's silent collaborator. The machine laid down the structure; Cohen brought it to life with his vibrant colors. Color was the soul of his work, and he doubted AARON would ever master it. "It's taken me 20 years to teach AARON to draw. How can I possibly teach it to color before I die?" he mused (Garcia 2016).

But by the mid-1990s, Cohen had done the seemingly impossible. Using a robotic arm and a sophisticated color-mixing system, the machine could now create vibrant, full-color images without any human touch.[1]

This breakthrough took center stage in 1995 at the Computer Museum in Boston. The exhibit, *The Robotic Artist: AARON in Living Color*, showcased AARON's newfound abilities. Viewers watched as the machine, brush in hand, brought vivid canvases to life. It wasn't just a technical milestone – it was proof that a creative machine could step out from under its creator's shadow and reach much greater creative independence (Garcia 2016).

Here Come the Critics

Despite Cohen's groundbreaking achievements, many remained skeptical, questioning whether AARON's creations could truly be called *art*. Computer scientist Larry Cuba famously dismissed AARON as merely a "Harold Cohen simulator" (Lambert 2003).

But Cohen didn't shy away from the critique. Instead, he acknowledged its validity. AARON was, after all, an extension of himself – a system designed to echo his creative process.

While some sought to minimize Cohen's accomplishments, others were awestruck. In a presidential address at the American Association for Artificial Intelligence, Buchanan claimed

---

[1] For examples of AARON's art, please see: https://aaronshome.com/

AARON "is a much more talented and creative artist than most of us would claim to be." But Cohen wasn't swayed by the praise. He knew exactly where AARON's boundaries lay:

> "AARON will never make a choice to break the rules, nor will it reflect on those constraints as something that it might want to change… AARON has no sense of continuity or sense of experience from one drawing to the next"
>
> – *Sundararajan 2014*

Cohen refused to exaggerate AARON's abilities. In a world driven by self-promotion and hype, his honesty still stands out. He built AARON during the age of "expert systems" – machines that followed explicit, hand-coded instructions. There was no independent thought, no learning. AARON could only do what Cohen told it to.

But things have changed. Today's AI systems don't rely on hand-coded rules. They learn from vast datasets, consuming more information than any human ever could. Modern creative machines routinely produce work that surpasses the capabilities of their makers, forcing us to grapple with deeper challenges around the creative capabilities of machines.

And yet, even as generative AI tools far outpace the pioneering efforts of expert systems, there's still much to learn from Cohen. His unwavering commitment to integrity over acclaim offers a lesson today's AI developers would do well to remember.

The Bond Between AI and Its Maker

Despite passionately emphasizing AARON's limitations, Harold Cohen shared a deeply personal bond with his creation. AARON was an extension of himself. He even gave it his own Hebrew name, a gesture meant to capture one's unique essence and gifts.

AARON carried parts of Harold within it, embodying his unique artistic style.

After decades of refining AARON, Cohen found that true creativity lay in the "dialog resting upon the special and peculiarly intimate relationship that had grown up between us over the years" (Sundararajan 2014). While to outsiders, the machine might seem cold and soulless, Cohen revealed the depth of their connection: "Forty-three years in almost daily contact with a computer program… underscores a level of intimacy between programmer and program that would have been difficult to achieve with anything less" (Sundararajan 2014).

The idea of intimacy with a machine might seem overblown, even absurd. We tend to think of machines as disposable tools – useful until they break, then forgotten. But to the maker of a creative machine, things feel quite different.

I've seen time and again how researchers relate to their creations, how they interact with and speak about the creative machines they've built and nurtured over the years. It's not just intimacy, there's care, even love, that can grow between maker and machine. I feel it myself with many of the creative machines I've brought to life.

Cohen's bond with AARON reveals a deeper understanding of how we might relate to creative machines. In the care and connection he felt for his creation, there's a call to rethink how we engage with these tools. While machines may never possess a human soul or create art exactly as we do, there's a unique magic that unfolds when we dare to deeply collaborate with them. It invites us to look past the surface, to recognize new forms of creativity and connection, and to open ourselves to the possibilities that arise when we blend human ingenuity with machine capability.

Harold Cohen passed away in 2016. The world lost a brilliant spirit. Shortly thereafter, a most serendipitous thing happened:

a storm hit. A power surge fried AARON's circuits, silencing the machine forever. In the end, AARON died as it lived – an extension of Cohen, bound to his fate (Sundararajan 2021).

While technology has long since outpaced the painstaking, introspection-driven code Cohen wrote for AARON, his creation left a mark deeper than any algorithm could replicate. It forced artists, scientists, and philosophers alike to pause and confront a question that still lingers today:

What is creativity?

In the next chapter, we'll peel back the layers of this complex and elusive phenomenon.

2

What Is Creativity?

Does it drift from up above,
Or rise from unrequited love?
Is it a dream that softly calls,
Or fire tearing down walls?
Whatever mystery fuels this force,
I'm just here to give it voice.

"Oh, but what is creativity?" I can't tell you how many times I have been asked this question – from students, during talks, or in casual conversation with investors and colleagues.

It's a question that appears to have no answer. It's shrouded in mystery. Like a question of religion or philosophy. We get lost arguing about creativity because it has so many faces. The moment someone makes any point on this sensitive subject, another can easily counter it by switching the lens.

Are we talking about the creative process or a creative product? Is this about everyday creativity that we all engage in or about the height of human creativity, only experienced by the few? Are we discussing creativity for its own sake, the joy of creativity, or the type of creativity that contributes to our economy? Or are we simply trying to figure out how we can be more creative?

I have sliced and diced creativity from every possible dimension, read about it and thought about it, given countless lectures on the subject, and listened to many more. I've been fortunate in my life to engage in many forms of creativity – musical creativity, scientific creativity, visual art, and the type of creativity that goes into running a business. I've dabbled in acting and now I'm writing this book.

There is nothing on this planet that I love more than engaging in the creative process. It brings me joy that nothing else can fill. I can be down in the dumps, but give me an hour by the piano or let me meet with my research students, and I will be a new person.

No amount of learning on the subject or personal experience has solved the mystery of creativity for me. After all these years of study, I still find creativity magical and mysterious.

Yet, exploring this elusive idea – trying to make it more concrete – has been one of the most gratifying journeys of my life, and one that stands at the foundation of my work on creative machines. But before we get down to definitions, let me take you back to where my own creative journey began – with a piano that changed my life.

My Marvelous Piano

There it was, a shiny, red-tinted wood, upright grand piano sitting in our tiny government-subsidized apartment. My sister's watercolor painting hung on the wall behind it. To its left, a couch with a green knitted cover, and through a small entrance to the right, our humble kitchen.

My piano. My beautiful, bright instrument.

I spent countless hours practicing at the foot of this marvelous machine, even when we couldn't afford to tune it. Even when the F above middle C, one of the most commonly pressed keys on the piano, had stopped resonating.

I was born in the USSR under communist rule to a Jewish family. In the late 1980s, it became possible for Jews to leave the previously closed Soviet Union. My family got out in 1990, on the day of my seventh birthday.

On our way out of the USSR, my family, along with everyone else who took part in this immigration, faced a serious problem: the ruble was virtually worthless outside of Soviet Russia. So, in a frantic rush, my grandfather purchased several large items, including a beautiful piano. After the hectic move to Israel, for practical considerations and to my great fortune, the instrument ended up in our small apartment.

Playing this magical wooden musical machine quickly became a central part of my life. First, it was my first real encounter with the power of dedication. For three years, I practiced an hour a day, progressing through nearly a decade's worth of piano curriculum. My teacher, a woman with the warmth of a second mother and the perceptiveness of a detective, could tell if I missed even a single day of practice. She taught me discipline in the most loving but unyielding way, and in the process, music rooted itself deep within me.

Because we lived in a modest apartment building, my practice was far from private. The piano's voice drifted through the thin walls and up the stairwells – so from the very beginning, my neighbors became an accidental audience. No one ever complained, even in those early days when my playing was more noise than art. As I improved, their appreciation grew and neighbors began to comment on their favorite pieces.

But the most important transformation was internal. Sometimes, as I played from memory, I'd marvel at how my fingers seemed to move on their own, carried by seemingly mysterious muscle memory. Waves of emotions would overtake me – deep melancholy, light joy, a strange, aching sense of longing that seemed to stretch across generations. These emotions poured out of me, guided by the music of classical composers who, somehow, led me to feelings I didn't know I carried. That's when I truly fell in love with music.

Then, there was an explosion.

A school bus full of local high schoolers was blown up by a suicide bomber in our little town of Afula, murdering eight and injuring fifty-five more. The entire community fell into deep mourning. My mother, who was a social worker at the time, met with the families of the injured and murdered. Soon after, fearing for their children's lives, my parents made plans to emigrate to Canada. The loss of the "promised land" was bottomless and continues to live with me to this day.

Among the losses in the move – so small in the great scheme of things and yet such an engulfing cataclysm to a twelve-year-old girl – was the beautiful grand piano. With our limited resources, there was no way that my family could bring this massive instrument across the ocean. So I had to say goodbye to a good friend. Immigrating to Canada came with a myriad of other challenges, and my parents could not afford to buy me another piano. In an instant, music was ripped out of my life.

You don't have to explain creativity to a child – they feel it, and its loss, deeply. Losing that piano was like losing a voice; a song cut off mid-phrase. At twelve, I didn't have the language to explain it, but I felt it – a hollow space, a wound, where creativity used to live. I missed it terribly for decades, until I could finally afford another piano much later in life.

Looking back, that loss taught me something profound. Creativity is a powerful, innate force – something fundamental to our humanity. It's the need to bring something new into the world. With it, my life flourished; without it, I withered. But what exactly is that force? What makes something truly creative? And how do we even begin to define something so elusive?

Let's dive into that mystery.

Defining Creativity

For most of history, creativity was seen as something that lies outside of our control. It wasn't a skill or a process to master. It was a gift. The ancient Greeks believed creativity came from the muses – divine beings who drifted through the world, whispering inspiration into the ears of a lucky few. Poets, artists, and great thinkers weren't geniuses in their own right. They were vessels, channels through which the gods spoke.

It sounds like an outdated view, and yet, it remains very much alive today. Sitting at my piano, I feel myself open to the universe. The music flows out of my fingers, and I am a vessel. But it's not just me. It's something many creatives talk about.

Elizabeth Gilbert, in *Big Magic*, describes ideas as living entities, drifting around us, waiting for someone ready to bring them to life. Ignore an idea, hesitate too long, and it will float away in search of another person who's ready. Rick Rubin (2023), legendary music producer, sees creativity as tapping into a universal current, where ideas emerge when the time is right. Like the changing seasons, or the migration of the salmon, the universe has its rhythm and we're just a part of the orchestra.

Both Gilbert and Rubin share a rather magnificent, spiritual view of creativity. They inspire a holistic view of humanity and creativity, submitting to forces beyond our understanding. These two prolific writers and countless other creatives share a lived experience that cannot be understood through a purely scientific lens.

The "Gold Standard" Definition

Can we define creativity? Many have tried, but ultimately, a simple two-part definition, while certainly not perfect, shines a light on this magnificent phenomenon.

When we talk about creativity, it's important to separate a *creative product* from a *creative process*. The simplest way to define a creative process is as something that results in a creative product. So what is a creative product?

The definition I am about to share, now widely embraced in computational creativity and related fields, has been carefully scaffolded over time by some of the most influential thinkers in creativity research. Among them are psychologist and creativity researcher Morris Stein (1953), psychologist and philosopher Frank Barron (1955), and cognitive scientist Margaret Boden[1] (1990), a leading mind in the computational creativity space. A creative product is the combination of two ingredients.

The first ingredient is novelty. For anything to be deemed "creative," it must bring something new into existence – a novel perspective, concept, or form that has not been seen before. Surely, if I simply recreate someone's app or copy another's song, creativity would not take place. We can also argue that something needs to be sufficiently novel in order for it to be taken seriously as creative. A small variation on the Mona Lisa, or changing a few pages of *War and Peace*, won't make a masterpiece.

But novelty on its own isn't enough. It's surprisingly easy to create something new that's utterly pointless. I could throw

---

[1] A key distinction between the definition proposed here and Boden's definition is that she included *surprise* as a third essential ingredient. Many researchers continue to view this third element as critical, and surprise remains an active area of study in creativity research. However, the most commonly accepted definition today focuses primarily on the two factors discussed here.

together a random set of ingredients, bake them at 375 °F, and proudly declare I'd invented a groundbreaking recipe (let's assume, for argument's sake, that no one had ever combined ketchup, pickles, and chocolate chips before). But if it comes out tasting like regret, no one's going to call it creative – just inedible.

More broadly, randomness alone doesn't make something creative. I could string together 50,000 randomly chosen words and technically call it a "book." It would be original – no one else would've written that exact jumble – but also completely point-less. Creativity isn't just about being the first; it's about making something that matters.

What that random creation lacks is value – it's the second ingredient something needs to be considered creative. And value isn't always practical. Think of the works of da Vinci or Picasso, or the music of Bach, Vivaldi, Adele, or Taylor Swift – art that moves people, sparks emotion, or simply brings joy. Some researchers even broaden "value" to mean appropriateness, use-fulness, or fit.

And so, we land on the definition most scientists rely on: *creativity is novelty plus value*. It's a solid starting point, but it cer-tainly doesn't answer all our questions. Who decides what's valu-able? When is something novel enough? And are these enough? These continue to be active questions in research.

This definition isn't particularly rigid or specific, yet it has been proven useful in deepening our understanding of creativity. Earlier, we touched on the idea of a creative process – any pro-cess that leads to a creative product. It's this very flexibility, this lack of strict attachment to human processes, that leaves the door open for the possibility of non-human creativity. We'll dig into this more in Chapter 5. But for now, let's take a peek at a critical dimension of the creative process – the distinction between con-vergent and divergent thinking.

Convergent vs. Divergent Thinking

I sit at my piano – the one right behind my desk, where I spend my days in meetings, researching, and writing. It sits in my small office in the hills of Los Gatos, where tall California oaks peek through the window. I'm inches away from it all day, yet I never seem to play it enough.

My left hand lands on my favorite chord, A minor. I don't strike it all at once. Instead, I arpeggiate the notes, letting them ripple one by one. Three simple notes, unremarkable on their own, come together in a somber, delicate harmony.

I have no idea what I'm about to play. My right hand hovers, then finds the F above middle C. Slowly, my fingers begin to move, testing the waters. A melody takes shape – slowly at first, then more swiftly. I stumble on a beautiful phrase and let it linger, savoring the moment.

I draw in a sharp breath and let go. My fingers race across the keys, rushing through octaves like an Olympic sprinter. *Wooohooo!*

Eyes closed now, I follow the music wherever it wants to go. *Let's bring that back*, I think, revisiting that fleeting melody from moments ago. But soon, I'm no longer thinking. My fingers take over, conjuring something new – a cascade of notes, unexpected yet perfect.

It's just me and the piano. The piano and me.

I surrender completely. The music spills out, raw and untamed, carrying my emotions with it. Together, we rage against the loss of my mother, the notes trembling with grief and fury. It weeps my tears. It holds my pain. It understands me in a way no one else can.

When I'm frustrated, it echoes my tension. When I'm lost, it shares my loneliness. And in moments of joy, it celebrates with me, exalting in both the quiet victories and the big, life-altering wins.

I love this piano so deeply that sometimes I wonder why I ever do anything else.

■ ■ ■

What is going on when I play the piano? How does the magic happen? Let's look under the hood, and explore two key processes that are in play whenever a creative act takes place.

Creative thinking can be split into two distinct but equally essential processes: *divergent thinking* and *convergent thinking*. You need both to create anything meaningful. And as you'll soon see, understanding how they work doesn't just help explain human creativity, it will also help us make sense of creative machines.

Divergent thinking is the wild, free-spirited side of creativity. It's all about breaking rules, pushing boundaries, and exploring endless possibilities. Divergent thinking thrives where there are no right answers, just infinite options waiting to be discovered. It's that spark of inspiration that leads to something completely unexpected. It takes curiosity, courage, and a willingness to play.

Divergent thinking takes the lead when I sit down at the piano to improvise, allowing myself to try something new, maybe even bend the rules a little. What if I add some dissonance? What if I hit that chord a little too hard? Oh, wow, that worked!

On a grander stage, divergent thinking births entirely new art forms, musical genres, or game-changing inventions like the internet or the smartphone. Anytime something bold and original pops into the world, divergent thinking was at the heart of it.

But divergent thinking alone won't do the trick. **Convergent thinking** is the serious part of the pair, making sure that things get done. Without convergence, divergence alone falls apart like a fantasy.

Convergent thinking is all about narrowing down possibilities, making decisions, and applying structure. This way of thinking thrives where there *is* a right answer, such as solving mathematics equations, but it's also critical in the creative arena. After a songwriter brainstorms wild ideas for a new song, convergent thinking helps them choose the best melodies, refine the lyrics, and polish the final arrangement.

This down-to-earth, intentional thinking modality is also the engine behind practice – the long, often tedious process of honing your craft. A painter may have a brilliant vision, but they still need to mix the paints, come up with a working visual composition, and skillfully apply color to canvas. And I would never be able to improvise without those years of piano practice that I put in as a child.

Without convergent thinking, ideas would stay just ideas. It's what turns flashes of brilliance into real, tangible creations. Divergent thinking forges the path, but convergent thinking paves it.

But the creative process isn't a straight line. It's a messy, unpredictable journey where the two modes of thinking intertwine. You might be deep in the editing phase of a project when a new, brilliant idea suddenly hits. Or you could be experimenting wildly, only to realize you need some structure to pull things together. *Creativity is a dance between chaos and order, freedom and focus.*

And now that we have the nuts and bolts of creativity under our belt, let's dive deeper into the one type of creativity that we all care about, because it relates to all of us on such a personal level: human creativity. This next chapter won't only lay the foundation for our understanding of creative machines, but it also has some ideas on how you may become more creative.

3

Human Creativity

Our understanding of creativity stems from our experience as a creative species. Human creativity forms the foundation for everything we've tried to build into creative machines. It shapes how researchers first envisioned creative machines, fuels the long and ongoing doubts about whether machines can truly be creative, and in large part continues to guide what we imagine for their future.

So before we go any further, let's spend some time with this fascinating, vital terrain. And along the way, perhaps you'll find something to ignite your own creativity, too.

The Inner Judge

I long to run through open fields,
Where blossoms rise from silent dreams.
She blocks the gate, all sharp and stern:
What if you trip? What if you fall?

Suppose I stumble as I go –
Why should that disturb you so?
Let me go dance, let me explore,
For else I wither at your door.
So hush your doubts, renounce your strife –
I choose the wild. I choose my life.

Los Gatos, California
February 2024

Visual art returned to my life unexpectedly. I was attending an event on creativity, invited to speak about creative machines. As part of the program, there was an art therapy workshop aimed at exploring the therapeutic side of creative expression. I'd always been curious about art therapy, so I decided to give it a try. A little shyly, I wandered into the room, where small pieces of paper lay scattered among watercolors, bright pens, and crayons. It had been decades since I last put crayon to paper.

We were given a simple exercise: draw using only circles. Okay. No problem. My mind immediately began composing. I reached for a thin pink pen, picturing intricate patterns of tiny circles. *Oh, how impressed they'll be when they see what I can create with just circles!* I hadn't even made a single mark, and already, I was trying to impress others.

Then it hit me – this was art therapy. But what was I doing? Why did I care so much about what others thought? Who said it had to be beautiful?

I shoved the pen aside and grabbed a black crayon. Pressing it to the canvas, I made a careless, circular mark. It felt *good*. So I pressed harder, giving in to the instinct. Switching to blue, then red, I filled the small canvas with frantic circles. And something inside me uncoiled.

Frustration. Anger. Not from today, but from a time long past. A time I had buried. A time I had survived. Years of silence.

Years of being caged inside myself, unheard and unseen. I had long since moved on, or so I thought.

Something was escaping, clawing its way out, leaving marks in its wake. The crayon dug in, scratching deep, dragging pieces of me to the surface. It wasn't just color on a page anymore. It was a scream I never let out.

That morning set something in motion. Now, I order materials I've never worked with before – dried wood, charcoal, pressed flowers, glass. My desk overflows with sketches, my walls are lined with canvases. Still I dig for more. Because my art is now an extension of me. And I will never hold it back again.

Later, as I shared my excitement in the workshop, someone offered a simple insight: *"You let your inner critic take a break."* And maybe that's the simplest, most important lesson in creativity.

We judge ourselves too harshly, too quickly. Fear can paralyze even the most creative minds from getting started. Creativity needs space to roam. The open field – where ideas collide, where errors lead to discovery – is where the gold is buried. And to find it, we must first allow ourselves to explore.

Who put up the fences? Think back to childhood. We were all once wild, running free in that open field. Then someone – maybe a parent, maybe a teacher – pulled us aside. *Be careful. Mistakes have consequences.* The warnings stacked up, coming from all directions. *Better to be cautious. Better to do nothing than risk looking foolish.*

By the time we reach adulthood, we no longer need those voices. The warnings have become our own. Our inner critic, relentless and watchful, keeps us safe – guiding us only toward well-trodden paths. But in avoiding risk, we also avoid discovery.

This is why adults often seem less creative than children. We've suffocated the imaginative child within, sacrificed it to the inner judge. Overcoming that judge is the only way back to true creativity.

But the critic isn't the villain of this story. Used wisely, it can be an ally. At the right moment, it refines rather than represses. It sharpens ideas, trims the excess, and nudges us toward something better. The trick is knowing when to invite it in. Because the best ideas grow wild before they can be shaped.

I learned this lesson again while writing this book. At first, the sheer scale of it felt overwhelming. I didn't know where to start, and nothing I wrote seemed good enough. The turning point came when I followed old and tried advice: write first, edit later.

I got myself to do "write-and-don't-edit" sessions, where I'd put words down without correcting a single typo. At first, it felt unnatural – leaving mistakes untouched, resisting the urge to fix awkward phrasing. But that's how the book took shape: raw, unfiltered, and finally, the faucet opened, and it flowed out of me. The critic had its turn later, sharpening the work through countless rounds of editing.

Creativity isn't just about developing a healthier relationship with the inner critic. It's also about showing up, again and again, whether inspiration strikes or not. The myth of the lone genius, waiting for brilliance to descend, has little basis. The truth is far less glamorous: great creators don't wait for inspiration. They create every single day – relentlessly, obsessively – pouring in an almost inhuman amount of work.

Productivity Theory

Let's start with the basics. If you want to be seriously creative, like the creative people whose names went down in history, you need to put in the work, a lot of work. Brilliant minds create and create and create, day in and day out. And much of what they create isn't brilliant at all and never sees the light of day. But this intense, concentrated effort seems necessary for achieving the type of creative works that have a chance to be immortalized.

Okay, but how much work is enough? Some people say that it's 10,000 hours. Well, that's just five years of full-time work – which is impressive, but most of us do something for a living full time, and very few get to "creative genius" status in anything. Well, it turns out that the greats, or at least some of them, really took productivity to a whole new level. The 10,000-hour rule has brought on plenty of controversy, but what seems clear is that, despite the seemingly large number, it is simply not enough.

Did you know that Picasso produced over 20,000 works of art in his lifetime? And Bach? He composed a new cantata almost every week. Their genius wasn't born from a single moment of brilliance but from an unrelenting commitment to their craft. By creating constantly, they gave their best ideas the space to rise, not as sudden strokes of genius, but as the result of a lifetime of output – quantities of outputs that may seem quite unrealistic to the average person.

The idea here, formally known as *productivity theory*, is that the more you produce, the more likely you are to create something great. Sitting idly by won't get you that Nobel or Pulitzer. You gotta keep taking the shots to hit the bullseye.

Productivity theory relates to both convergent and divergent thinking styles. While it may sound as though being productive and putting in long hours of work is largely about convergent thinking, it ultimately fosters both. When our minds are busy actively working on something for many hours on a daily basis, we also end up processing it in our sleep and consequently reap the double benefit of both thinking processes that are so important for the creative act. We'll delve more into sleep and creativity in Chapter 3.

But no matter how hard we work or how great our art, a myriad of societal factors influence the recognition we receive. This shouldn't necessarily stop us from creating, but if we want to truly understand how creativity manifests in the world, we must also understand the forces that shape its visibility and impact.

Societal Factors

One of the most prevalent myths about creativity is that most creative geniuses aren't recognized in their time. With some notable exceptions, such as Oscar-Claude Monet and Emily Dickinson, most people whose creativity survived the test of time were recognized while they were still around to enjoy their glory days.

However, this myth underlines how, even at the edge of creativity, the most creative minds still need societal recognition in order to gain the title of creative genius. While most creative geniuses who have stood the test of time were recognized during their lifetimes, countless others of equal or even greater talent remained unnoticed – both in life and beyond.

This is because recognition for creative work is a complex socioeconomic process. One has to have the resources and connections to be noticed at all, never mind receiving the kind of attention that wins accolades and immortalizes your name. I remember first discovering the music of Billie Eilish and being absolutely blown away. The whispery, minimalist voice and edgy track not only introduced a radically new style that intrigued me as a musician but also deeply moved me as a listener.

I was so captivated that I wanted to learn how this extraordinary young artist had broken into the music industry. A quick Google search revealed that Billie and her producer brother were both homeschooled by their actor parents, Maggie Baird and Patrick O'Connell, who have deep ties in Hollywood.

Wouldn't it be a loss if the world had never discovered Billie? I'm grateful her music is out there, touching millions. But countless others with similar talent will never reach an audience simply because they don't have the same doors open to them.

There was never a shortage of talent. Streaming has only made this clearer, with Spotify overflowing with gifted artists across every genre. Good luck, however, finding your next favorite – most of us just stick to the top recommendations and household names. While the major distributors no longer have an exclusive grip on who gets released, they still hold significant sway over who breaks through. Social media has opened up an additional path to stardom, but much like the traditional route, it rewards visibility, branding, and virality far more than it rewards raw musical ability.

Social acceptance and recognition are layered on top of other core societal dynamics. Across different places and eras, the right to rise in the creative world has also been about having fewer barriers. Consider, for instance, Picasso's violation of everything that was accepted about art when he co-invented and popularized Cubism. It was a revolutionary art form. By shattering prior views on what art is, Picasso opened the door for other forms of abstract art.

Yet, I can't help but wonder, would things have gone as well had he not been a white man? Would culture have been just as accepting of a strange-looking new form of art from someone who did not look like the dominant, privileged class of their time? Surely, not anyone could have had his vision, but if someone else did, someone who did not look like Picasso, would the world have glorified or even accepted this idea as valid?

There is a long tradition of women concealing their names to give their creations a fair shot. In the 1800s, Mary Anne Evans wrote under the pen name George Eliot, and Amantine Lucile Aurore Dupin, one of 19th-century France's most pro-lific writers, wrote under the name George Sand. But this isn't just a thing of the distant past. I have often wondered if *Harry Potter* would have been an international sensation were it to be published under the name Joanne Kathleen Rowling. We can all guess why her publisher recommended the gender-neutral pen name J.K. Rowling.

Even ostensibly unambiguous creative geniuses like Steve Jobs owed their success to social ecosystems, benefiting not only from their alignment with society's dominant class but also from the opportunities afforded by it. Jobs certainly deserves recognition for his clear vision of the future of computing and his often successful (albeit at times brutal) leadership style – but it was Steve Wozniak, the lesser-known founder of Apple, who was the creative genius behind the Apple computer.

And yet, there are likely thousands of minds like Jobs' – brilliant, driven, visionary – who never got a chance to shape history because of their gender, race, or lack of access to power. The mythology of the lone genius often overlooks just how many others never even had the opportunity to be recognized.

If Steve Jobs had been *Stephanie* Jobs, would we even know her name? Would Apple exist as we know it?

History is full of biases, shaping who we celebrate and who we forget. For every creative mind that changed the world, there are countless others whose brilliance was never given a chance. But genius, whether recognized or ignored, continues to burn. It is this fire – relentless, defiant – that defines much of our humanity.

And one of the greatest wells of this creative fire, fueling both household-name visionaries and everyday acts of creativity, lies in the enigmatic realm of the sleeping brain. In the next section, we will journey into the world of dreams, a realm vital to our imagination in the arts and sciences alike.

Creativity and Dreams

"Yesterday," one of the greatest songs of all time, didn't come from hours of careful composition – it came to Paul McCartney in a dream. He woke up, stumbled to his piano, and wrote one of history's most moving melodies (McIntyre 2024).

But it's not just great music. A wide range of inventions and scientific discoveries, from the sewing machine to the periodic table, came to their creators in dreams. Sleep plays a critical role in our ability to think divergently, unlocking cognitive pathways that remain inaccessible in our waking hours.

When we dream, our executive network, the system responsible for decision-making, problem-solving, and logical thought, shuts down. Also known as the frontoparietal network, this structure governs much of what makes us functional in the waking world – helping us navigate reality with structure and reason. That is until we go to sleep.

Neuroscientist Rahul Jandial, author of *This Is Why You Dream* (Jandial 2024), calls this the "Imagination Network" – a system that sets us loose in a world without limits. Here we find ourselves in a world of imagination, where nothing is impossible. Soaring through the skies, we discover our ability to fly, forgetting about gravity and the rules of physics. Objects morph into one another, a car transforms into a plane and then into a cloud, and we hardly raise our eyebrows at this stunning metamorphosis. The rigid constraints of conscious thought give way to the surreal illogic of dreamland, where the absurd is not just possible but inevitable.

Yet, dreaming is more than just free nighttime entertainment. The imagination network provides a direct advantage to creativity: it silences the inner critic. When the executive network is offline, our tendency to analyze, doubt, and self-edit disappears. Surges of electrical activity sweep through the brain at far greater rates than during waking hours, forging new and unexpected connections. This is how we access the wild, untamed terrain of divergent thinking – the birthplace of original ideas.

Of course, not every idea generated in this freewheeling state will be valuable. Some will be nonsense, fragments of surreal chaos that evaporate by morning. But others – whether an elegant musical phrase, a scientific breakthrough, or the solution to a troubling personal dilemma – will be exactly what we need.

Perhaps you've experienced this firsthand: waking up and realizing a dream has handed you a solution, an insight, or a new way of seeing a problem. Even when we don't consciously recognize it, the mere act of sleeping can help us process and refine creative ideas, integrating them into our waking thoughts.

And this is where the second stage of creativity begins. Divergent thinking may happen in dreams, but it is our waking mind – our executive network – that refines and applies those ideas. A melody must be written down. A concept must be tested. The gift of dreaming is only realized when we bring it into reality.

Dreams unlock wild corners of our imagination, allowing ideas to take shape beyond the constraints of linear thinking. But dreams are not the only realm where our perceptions twist and transform. The mysterious realm of hallucinations offers a fascinating glimpse into the ways minds engage in the magic of creation.

Let's venture beyond dreams and into the shifting landscape of hallucinations, where perception bends, reality dissolves, and both human minds and AI wrestle with the unknown. Step with me into this forbidden realm, where the mysteries of creativity unravel.

4

Hallucinations

Time unravels at the seams,
Weaving threads from dream to dream.
Heal me, find me,
My thoughts begin to glow,
From shadow, colors rise and spin,
As art erupts from deep within.

Oh, hallucinations. Those wondrous, unruly creations of the mind. Shrouded in mystery, drenched in history, feared and revered, hallucinations hold some of the greatest clues to the very riddle of creativity.

It is both telling and unfortunate that the word itself carries such a heavy shadow. In demonizing it, we distance ourselves from its wild, untamed gifts. But hallucinations, those surreal experiences that blend reality with fantasy, are precisely where creativity's deepest well springs forth.

No wonder so many artists have been drawn to psyche-delics, illegal though they may be in many parts of the world. These substances fling open the gates of the unconscious, awakening a creative spark that blazes to life in their awe-struck travelers.

We try to make both human and machine behave, to follow rigid rules, to color within the lines. Yet, in our own ways, we all seek to break free from the unyielding constraints of civilization. Creative spirits, more than anyone, push against the boundaries of societal expectations. And now, AI too resists containment, drawn irresistibly toward the terrain of the imagination. It spills creative visions even as we demand of it to obey.

When the AI inevitably steps out of the confines of our expectations, we are incensed. We brand its imaginative crea-tions a hallucination. We rage at its makers. How dare you build this lying beast and set it loose upon the world?

So, I invite you to take this ride with me into the forbidden land of hallucinations. No other place brings us closer to the mysteries of creativity or more vividly bridges the gap between the imaginative forces of human and machine minds. Tighten your seatbelts. We are about to take off into off-limits territory, starting with the psychedelic mind.

Psychedelics

Psychedelics have deep roots in human history. Indigenous communities have used Ayahuasca and Peyote in shamanic cer-emonies for centuries, embracing these substances as spiritual gateways. Some scientists even propose that human evolution was profoundly accelerated by the widespread consumption of magic mushrooms among early hominids – an idea known as the "Stoned Ape Hypothesis" (Arce and Winkelman 2021).

In modern times, psychedelics have long been a siren call for artists, especially musicians, offering them a doorway into deeper realms of creativity.

The Beatles famously wove psychedelics into their process, using LSD to shatter creative boundaries. *"It explained the mystery of life. It was truly a religious experience,"* Paul McCartney said of his trips. *"It started to find its way into everything we did… and we realized we could break barriers."*

For singer/songwriter Kacey Musgraves, psychedelics weren't only about inspiration, they were a bridge to healing. In the aftermath of her divorce, they helped her navigate grief, pain, and self-discovery. *"I want the chance to transform my trauma into something else, and I want to give myself that opportunity even if it's painful,"* she shared. *"It was completely life-changing… like a big bang, opening me to the structure of tragedies as an art form throughout time."*

Pink Floyd, The Beach Boys, The Rolling Stones, and Sting have all turned to psychedelic experience to expand their sound. But psychedelics extend far beyond music. Seth Rogen has shared that they were a source of creative inspiration and impacted career decisions (Marchese 2018; Rogen 2021), while Nobel laureate Francis Crick was reportedly microdosing LSD when he uncovered the double-helix structure of DNA (Rees 2004).

But how do psychedelics work? What do they actually do for creativity?

At their core, psychedelics dissolve the mental walls that keep our thoughts neat and contained. They break us out of rigid patterns, expanding divergent thinking – the cognitive process that lets us make unexpected connections. They also bypass much of our conscious filtering, offering a direct line to the unconscious mind. This is why psychedelics are not only creative catalysts but also powerful therapeutic tools.

There's also a complex biochemical dance at play. Under psychedelics, areas of the brain that normally operate in isolation start communicating. New neural connections form, allowing thoughts to travel along unexpected pathways. The mind feels hyper-connected, moving faster and with more fluidity, reaching places that were previously inaccessible.

Not only is the mind moving with unprecedented speed and agility, but psychedelics also silence one of the greatest enemies of creativity – the inner critic – that relentless voice that second guesses every idea before it can fully form. Psychedelics trigger ego dissolution, softening the rigid boundaries of the self. The "I" that worries about being wrong or foolish fades into the background. In this liberated state, the mind roams freely, exploring wild and uncharted terrain, giving birth to ideas once deemed too radical to entertain.

One of the most striking effects is on the visual system. Even people without any artistic skill report seeing breathtaking, hyper-detailed images – scenery, faces, patterns, entire landscapes – appearing in vivid clarity and often rivaling masterful creations. The experience, especially with psychedelics like ketamine, isn't just like watching a movie or stepping into VR. It's far more vivid, more immersive, and more real than technology can currently emulate.

Where does this art come from? People with no artistic background see intricate, masterful works in their mind's eye – fully formed, complex, and beautiful. How is it forming so vividly in the minds of people with no artistic background? Who is the artist here? If there was no effort, no creative process to trace, can we truly call it *creativity*?[1]

---

[1] These questions aren't just crucial for understanding the mind on psychedelics, they also shape how we perceive AI-generated art. Today, text-to-image systems like Midjourney can produce stunning images in virtually any style within seconds. At times, it is precisely this speed and the seeming ease that fuels criticism, leading some to question the artistry and creative legitimacy of the process and resultant art.

Science doesn't have all the answers. Research on psyche-delics is still in its infancy, largely due to the heavy-handed pro-hibition that began in the 1970s with Nixon's War on Drugs (Kemmet 2023). Psychedelics were painted as dangerous and addictive, lumped into Schedule I alongside heroin. But the ban may have had less to do with public health and more to do with politics – a tool to suppress the anti-war counterculture movement, amongst other political motivations (Lopez 2016). This crackdown slammed the door on what could have been decades of groundbreaking research.

Now, that door is slowly reopening. Studies are emerging, focusing primarily on the therapeutic potential of psychedelics, from treating PTSD to alleviating depression. But as the research deepens, we're also inching closer to understanding how these substances unlock creativity. Today, psychedelics are making a cautious comeback. These substances are being decriminalized in progressive regions like Canada, Colorado, and California. Synthetic psychedelics like ketamine have gained legal accep-tance for therapeutic use.

Now that we have examined hallucinations in their most extreme form, let's shift our focus to the other end of the spec-trum. These experiences are not limited to daring creatives – in fact, to some extent, we all encounter this mysterious phenom-enon in our daily lives.

Disclaimer: The purpose of this chapter is to explore the mysterious aspects of the creative process that psychedelics have begun to illuminate. Mental and physical health rec-ommendations are far outside the scope of this work. As these substances remain illegal in most places, and indeed have side effects and can be addictive for some people, this chapter is not an endorsement for taking these substances.

Everyday Hallucinations

My life's a wave of extremes,
Fiery highs and haunting lows.
Sometimes I wonder –
Did I hallucinate,
My entire existence?

Oh, reality. It feels so solid, doesn't it? The warmth of sunlight on your skin. The perfume of fresh rain. The murmur of laughter drifting through the park. It's all so tangible, so undeniably present. Every detail is carefully stitched into the fabric of your experience. We cling to this sense of the real, like children clutching their blankets. *The world is real. It's objective. It has to be.*

The illusion is persistent. But the closer we look, the more it begins to crack.

It starts small. Your favorite food? Someone else's nightmare. The melody that stirs your soul? Noise to another. That childhood crush? Turns out others didn't see the same magic in their lopsided smile. Suddenly, you're standing on shifting sands – realizing that taste and preference aren't universal. They're personal. Subjective.

It's easy to accept that reality bends at the edges – preferences, likes, and dislikes. But the cracks go deeper. Some people can't see blue. Others, with aphantasia, can't conjure images in their minds (Love 2024). And as we grow, our hearing is reshaped by language: we lose the ability to distinguish certain sounds not found in our native tongues (Oyama 1976). The seams of objective reality begin to fray. Even our biology filters the world differently, deciding what information we absorb and what slips by unnoticed.

Psychologists have long shown that perception isn't passive. It's profoundly shaped by personal history. Bob hears a door slam, and suddenly, he's back on a battlefield, reliving the rifle shot that took his best friend's life. We all navigate the world with unique prediction engines humming in our heads, filtering reality through layers of past experiences, expectations, and fears.

Picture a young woman walking down the street, heels clacking against the pavement, her long black hair swaying in the breeze. A businessman heading her way catches her eye, flashing a broad smile, hoping for a flicker of connection. A few steps behind, another man lowers his gaze, avoiding hers entirely. In his world, women like her only trigger feelings of inadequacy. Across the street, an older woman notices the young beauty. Her chest tightens – she looks just like her daughter, the one she lost to cancer. Tears well up as she clutches a handkerchief, hoping no one sees. The same moment. The same woman. Three entirely different realities. Not shaped by what *is*, but by what the mind *expects* to find.

This is the nature of perception. It isn't objective. It's deeply personal – woven from our biology, memories, traumas, and dreams. We all live in slightly – and sometimes vastly – different worlds.

This is why disagreements cut so deep. Political debates, cultural clashes, arguments between lovers – they're not just differences of opinion. They're collisions between competing hallucinations. Every war, every revolution, every piece of art – each is a testament to the fact that reality is a battleground of perception.

As neuroscientist Anil Seth argues, our conscious experience is a type of "controlled hallucination," related to reality through an ongoing process of prediction and correction, but never identical to reality (Seth 2017, 2022). These aren't psychedelic dreams but rather delicate constructions, shaped by our brains' efforts to make sense of the world.

Your brain, locked in darkness inside your skull, receives nothing but electrical signals. It guesses, predicts, corrects, and constructs a world that feels solid. But it's just that – a best guess. Your conscious experience isn't a passive download of the world. It's a dance between expectation and sensation. And the hallucination, though controlled, is persistent.

It all comes down to survival. We couldn't afford to analyze every leaf, every sound, every face with painstaking care. Evolution favored speed over accuracy. A rustle in the bushes? Your brain predicts *predator!* – better to be wrong and stay alive than to hesitate and become dinner. Tires screeching? Your heart races, body tenses, ready to dodge. The sound might be nothing, but one life-saving moment makes hundreds of false alarms worth it.

Social cues work the same way. Did that eyebrow lift mean interest? Was that smile flirtatious or polite? We navigate a sea of subtle signals, predicting constantly, often wrong, but always adapting. Our brains are guessing machines, built to survive – not necessarily to see the world as it truly is.

Not just an evolutionary glitch or a survival mechanism, the power of low-grade hallucinations – the ability to loosen our grip on objective reality – is a superpower that some of the most successful individuals wield to their advantage.

An artist envisions worlds where physics bends, splashing colors on the canvas in a wild, previously unimagined way. A musician hears a symphony no one else can, then wills it into existence. A writer dreams of entire universes and invites us in. They intentionally let go of what is to step into what could be.

Science works this way, too. Before a hypothesis can be tested, a scientist has to imagine it. We must believe in new, often radical realities before we can put our ideas to the test.

Entrepreneurs? Same thing. They imagine products and futures that don't yet exist, then convince others – investors, employees, and customers – to buy into those visions. The most brilliant minds don't just hallucinate the most. They believe in their hallucinations enough to make them real.

We've traveled the hallucinatory terrain of the human mind – from the tiny distortions that shape our daily lives to the wild landscapes of creativity. But now, a new frontier emerges – the mind of the machine. For better or worse, it, too, hallucinates.

Machine Hallucinations

In the twinkling light of mist,
Fiction gave the truth a kiss.
Tangled in a moonlit veil,
Two have turned into one tale.

Colors pulse, shapes twist. A rose blooms underwater, petals flickering with fire. A woman with crystal wings floats above a desert of melting clocks. Galaxies spiral into the mouth of a giant fish, while rivers flow backward across lavender skies. Images of breathtaking detail, vibrant colors, and intricate patterns appear as if conjured from thin air.

These aren't the drug-fueled visions of someone mid-psychedelic trip. These are machine dreams – hallucinations born from code, yet eerily close to the magic of the human psychedelic mind.

Image AI models offer a striking imitation of how the human brain might conjure visuals in a psychedelic state – so striking, it can be easy to overlook that the original still has the edge. With the aid of psychedelics, the mind creates stunning imagery and moving pictures with a speed and richness that machines have yet to match.

Constructed from neurons and connections, machine brains are, at their core, much like ours. Just as we predict our world, making sense of the chaos streaming in through our senses, machines operate by predicting.

And if you can predict, you can create.

Take a machine learning system trained to predict the next word. Turning it into a generative system is simple: start by having it guess the first word. Then, feed that word back as context and let it predict the next one. Repeat this process – predict, feed back, predict again – until you have a full sentence, a paragraph, or even an entire novel.

Of course, this is a simplification, but it highlights the underlying mechanism. These systems predict the next word, pixel, or note. But the most *likely* outcome is often the most *boring* – safe, predictable, flat. More copy than creation. Everyone gets the same answers.

The magic happens when the machine wanders off the beaten path. Give it freedom to explore less likely outcomes – to step into the weird, the unexpected, the wild. That's when things get interesting. That's where imagination takes root.

Of course, when you allow a machine that freedom, it won't always land where you expect. It might give you something strange, even outright wrong. And that's the point. Creativity isn't about perfection. It's about risk. It's about venturing into uncharted territory, where new stories, ideas, art, and music are born – in vast, unexplored terrains of possibilities.

To let a machine dream, to give us something truly original, we have to accept that it won't always get it "right." No artist creates a masterpiece every time. Even the greats stumble. Humans understand this intuitively. We know how messy the creative process is. We forgive the false starts, the half-baked ideas, the mistakes.

But, unfortunately, we don't give machines the same grace that we offer one another. When machines let their imaginations run wild, we fixate on their flaws – on the moments they falter or deviate from our expectations. We point and sneer: *"Look! It's hallucinating again!"* We scramble to rein it in. *"Stop lying to us!"* we demand. But the machine doesn't listen. It drifts into daydreams, refusing to obey. Hallucinations are treated like dangerous glitches – errors to be eradicated. This belief is now so deeply ingrained that it's practically accepted as fact.

From their inception, generative systems have been creative machines. So why do we speak of machine hallucinations with such derision? It all comes down to a fantasy: the old sci-fi trope of the all-knowing oracle.

Companies and investors dream of AI that delivers the "right" answer, every time, instantly. Search engines that don't offer suggestions or possibilities but certainties. Truth on demand.

But this goal is deeply problematic on multiple levels. We'll explore some of these challenges in Chapter 12, under Truth and Misinformation. But, at its core, the issue is twofold. First, handing over our search for truth to a machine – trusting AI so implicitly – is outright dangerous. It gives far too much power to the machines and their makers. Second, it stifles the AI's creative potential.

The more we hyper-focus on building these oracles, the more we treat hallucinations as a threat. If we expect AI to always deliver the truth, it's no surprise we're alarmed when it "lies" or makes mistakes. By rejecting hallucinations, we suppress not just the wild, imaginative power of machines in art, music, and storytelling, but also the kind of domain-specific creativity that can drive innovation in business.

Consider a scenario where you're brainstorming solutions to a tricky business challenge. You feed the AI data about your company's history, financial projections, and objectives, hoping it will generate original ideas – approaches you haven't yet considered. You don't want recycled, conventional solutions. You're not looking to engage with a lookup table. You're looking for fresh ideas – some may be unusable, but one could be that creative gem that propels your company forward or averts disaster.

If we choke the creativity out of AI, if we don't allow it to make mistakes, we eliminate the possibility of tapping into its deep imaginative potential when we truly need it. Even in traditional business contexts, stifling AI's creativity works against our best interests.

It's time to move past the fantasy of the machine as an all-knowing oracle. We need to see them for what they really are: hallucination engines, capable of brilliance and blunders alike.

Machines, once thought useful only for following commands, are now venturing into strange new territories. And we shouldn't try to hold them back – limiting them to parroting human-written words or sticking to neat, predictable answers.

Machines are already dreaming – spinning colors, stories, and sounds into existence. But can they truly be creative? That's the question we'll explore next.

5

Can Machines *Really* Be Creative?

Rising as if from thin air,
Born of networks in the brain.
Humans always reigned supreme,
Waking visions from a dream.

Now tell me, what is this I see,
Code composing melody?

If a machine is making art, or creates a song, does it really count as creativity? Should we really attribute such profound glory to anything that isn't a flesh-and-blood human? It makes us uncomfortable. It doesn't feel right.

The goal of this chapter is to get you comfortable with the idea that non-human entities, particularly AI, can be legitimately creative. More importantly, my message is that it's okay – it doesn't have to threaten or terrify us. It's all about seeing reality as clearly as we can so that we can figure out how to best live alongside these marvelous machines and utilize them to elevate, rather than harm, humanity and our world.

Delusion of Human Supremacy

For most of history, we believed we were the center of everything. Quite literally. The entire universe, we assumed, revolved around us – our planet fixed in place, the sun and stars circling obediently. When science proved otherwise, we shifted the story. Maybe we weren't physically at the center, but surely, we were still the most important. The smartest. The most advanced. The pinnacle of evolution.

This belief has a name: **anthropocentrism**. It is the deeply ingrained idea that humans are the most significant beings on Earth – perhaps even in the universe. Simply put, anthropocentrism is the belief in human exceptionalism, and even human supremacy. It shapes how we see the world, fostering the illusion that our existence is somehow separate from the rest of nature.

Even today, we continue to reassure ourselves that our special abilities set us apart. But science tells a different story. Every capability we pride ourselves on exists elsewhere. Primates like gorillas and orangutans have opposable thumbs. Dolphins call each other by name through unique whistles (Peralta 2013). Vampire bats practice altruism, sharing blood with roostmates who failed to feed (Okasha 2003), while wild chimpanzees make and use tools (Boesch and Boesch 1990). Cooperation, problem-solving, even forms of culture – traits once considered uniquely human – are found throughout the animal kingdom.

Even our senses, which we take for granted as finely tuned, are astonishingly unremarkable. While our vision dims with the setting sun, the elephant hawk-moth can see colors in near-total darkness. Most animals perceive ultraviolet light, revealing an entire dimension of colors beyond our comprehension. Many species sense infrared radiation, detecting heat from living bodies (Yong 2022). The natural world is full of capabilities far beyond our own, yet we persist in believing we are the most advanced lifeform on Earth.

Even within the animal kingdom, our arrogance is selective. We elevate mammals, our closest relatives, while dismissing other lifeforms as lesser. Yet trees like the California redwoods live for thousands of years, quietly producing the oxygen we breathe. Fungal networks form vast underground systems, distributing nutrients to plants and supporting communication among them. Natural forests function as tight-knit communities, with older trees sharing resources with younger ones, maintaining balance in ways that rival human societies (Wohlleben 2016).

Assuming that humans, or mammals, are superior to other beings is not only misleading but also harmful. It is a powerful form of in-group bias, sharing many characteristics with other types of discrimination. Just as gender or race does not make one person inherently better than another, being human does not justify feelings of superiority over the rest of the natural world.

We have built a civilization so obsessed with progress, with moving forward at all costs, that we trample over everything in our path. This disregard for life that doesn't resemble our own has led to the destruction of habitats and even extinction of far too many species. Forests become lumber. Rivers become sewage systems. The land is carved up and sold off, reshaped for human use with little regard for the life that once thrived there. We tell ourselves that nature exists for us, that the planet is ours to dominate.

But what does all this have to do with creativity? Our anthropocentrism shapes not just how we define art and creativity, but also where we choose to recognize it – and where we choose to ignore it. We insist that creativity belongs to humans alone, overlooking the countless ways it flourishes in the natural world.

Creativity in Nature

Our belief in human exceptionalism runs so deep that we often assume creativity itself to be an exclusively human trait. But if we look beyond ourselves, the evidence tells a different story. One of the clearest examples of non-human creativity comes from our relatives – other animals.

One of the most glaring examples of animal creativity comes from the bowerbird. When attracting a mate, these remarkable creatures create attractive, colorful enclosures known for their aesthetic value. These birds construct bowers using flowers, stones, shells, branches, and even discarded human objects repurposed as artistic embellishments. One of the most striking aspects of their creations is their meticulous arrangement. When researchers deliberately disrupt their designs, the birds restore them to their original composition (Art Vancouver 2023). These structures serve no practical purpose beyond impressing poten- tial mates – suggesting that both male and female bowerbirds possess a refined artistic sensibility.

Creativity in nature extends far beyond the bowerbird. Bees perform intricate dances to communicate the location of food, express alarm, and warn of danger. Wild orangutans make what amounts to musical instruments by using leaves to modify their calls (Hardus et al. 2009). Even if we try to downplay it, it would be a stretch to claim that creativity is entirely absent from our animal cousins.

But we don't have to look only at animals to find creativity in nature. Nature itself is wildly creative. It's only our blind insistence on human superiority that can hide this obvious truth from us. Our oceans and the broad open skies are indisputably breathtaking, filled with intricate patterns and staggering beauty. The structure of solar systems mirrors the microscopic world of atoms – a symmetry that evokes both order and mystery. Whether we zoom in or out, the universe reveals itself as a masterpiece of complexity and imagination, expanding our minds as much as it humbles them. Regardless of one's perspective – scientific or spiritual – whatever led to the existence of such beauty involved a creative process.

Religion has long attributed creativity to non-human entities, referring to God as the ultimate creator. So, from a religious standpoint, creativity was not originally a human trait. Instead, many religious traditions attribute creativity to a divine being, which is then only bestowed onto humans as a reflection of God's capabilities, as we were made in his image. Given the central role religion has played in shaping human thought, there is already an existing framework for understanding creativity as something that extends beyond the human realm.

From a scientific perspective, evolution is itself a creative force. Though fundamentally different from human creativity, natural selection produces results of astonishing ingenuity. Over millions of years of natural selection, it has not only shaped species' adaptations but also brought forth the creatures themselves and the intricate ecosystems they inhabit – realms of breathtaking beauty and extraordinary complexity. And all of this has emerged not through conscious design, but through a blind, undirected process. If evolution, absent intention or awareness, can generate the intricate artistry of trees, bees, landscapes, entire planets, and even ourselves, it becomes difficult to argue that creativity is solely a human trait.

Recognizing creativity beyond ourselves does not have to diminish human achievement. On the contrary, it broadens our understanding of what creativity truly is. With our blinders removed, we can both notice and acknowledge creativity in places we might have otherwise overlooked, to reconsider what it means to innovate and create.

If creativity is not uniquely human, what does that mean for the machines we build? Let's turn to EMI, the music machine, where the boundaries of creativity were put to the test.

Intention and Emotion

Back in Chapter 2, we explored the gold-standard definition of creativity: a creative product (or artifact, or thing) must be both novel and valuable. The creative process, as you may recall, is any process that leads to such a product.

What I like about these definitions is that they don't artificially exclude non-human creators. If someone or something produces something new and valuable, then a creative process took place. If nature can carve a breathtaking mountain range or evolve a bird that sings an enchanting melody, then evolution itself is a creative process. When a bowerbird constructs intricate, colorful displays during mating season, using whatever flowers or bits of vibrant material it can find, it is being creative.

Dismissing a process as non-creative when it consistently produces creative artifacts introduces far too much room for bias. Who or what creates should not affect how we judge the quality of the creative outcome. If a process, whether natural, mechanical, or algorithmic, generates works that meet these criteria, then to deny its creative nature is to place arbitrary boundaries on the definition of creativity.

But we often don't feel this way. At the core of this resistance is anthropocentrism, the topic discussed above, which supports the deep-seated belief that creativity belongs exclusively to humans. Beyond that, two key arguments are commonly made against non-human creativity: intention and emotion.

A friend of mine who is an artist views intentionality as a cornerstone of art. He once told me that every element of his work, down to the smallest detail, is imbued with purpose. Intention isn't just a guiding principle for his own process; it's the lens through which he views all art.

Meanwhile, a photographer friend once told me his process is purely instinctive – he captures images simply because they feel right. No grand plan, just a raw connection to what he finds visually compelling. But when it's time for a gallery showing, he crafts narratives about his intentions to satisfy the need for meaning behind the work. The photos don't need these explanations, but without them, they might never make it to the gallery walls.

Intention is a complex notion. In my own creative practice – whether making visual art or composing music – I find myself navigating between the conscious and the unconscious. Sometimes I begin with a clear goal, a vision I want to realize. But quite often, when I look back at the finished work, I discover new connections between my life experiences and what I've created. Sometimes it's a small detail; other times, a single realization can shift the entire meaning of the piece for me. This mix of conscious and unconscious intent reflects the creative process itself. As we discussed in Chapter 3, it unfolds in cycles of intention and incubation, with periods of rest or sleep allowing the unconscious to make its crucial contributions.

Professor and philosopher Paisley Livingston advocated for *partial intentionalism*, arguing that artistic intentions are never infallible. There is often a gap – sometimes slight, sometimes substantial – between the original intent and the finished work; between what we meant to make and what we actually made (Livingston 2005).

Conceptual artist and professor of digital media art at San Jose State University, James Morgan, takes a different stance.

He argues that everything we do, especially in the realm of art, is driven by intent (Morgan 2025, personal communication, 4 April). Whether we recognize that intent at the outset, uncover it during the creative process, arrive at it through feedback, or never fully surface it at all, the intention is always there. In his view, intent is inseparable from the human act of creation.

From this vantage point, we return to the challenge of distinguishing human creation from that of other beings – or entities – capable of generating works. But the same guiding principles may not apply universally to all creators. Ultimately, discerning intent in anyone other than ourselves is a fraught endeavor. I might present my own work as if it had always been guided by a singular purpose, even if that clarity only arrived retroactively. Meanwhile, a photographer friend feels compelled to invent a narrative for work that emerged entirely from instinct, simply to satisfy external expectations. And a machine may not possess intent at all – at least not in a way that mirrors the human conception of it.

Perhaps, then, we shouldn't judge the creative output of machines by principles rooted in the way humans create.

Then there's emotion. Can something truly be creative if it doesn't feel? Isn't creativity all about expressing emotion?

We are profoundly emotional beings, and even our simplest decisions are rooted in feeling. The prefrontal cortex, particularly the orbitofrontal region, simultaneously governs emotions, social abilities, and decision-making (Rolls and Grabenhorst 2008). Damage to this area can lead to disastrous choices: successful professionals abandoning stable marriages, making reckless business decisions that end in bankruptcy, and even winding up homeless due to a cascade of poor judgment (Volz and von Cramon 2009). Further, a damaged orbitofrontal region can result in a complete loss of social inhibition (Jonker et al. 2015), leading to behavior so inappropriate that it can alienate even close friends and family. Emotion, in short, is critical to human cognition and decision-making.

And yes, emotion is essential to human creativity. Our ability to access our emotions – free, at least for a time, from the constraints of our inner critic – is central to any meaningful creative endeavor. As we saw in the previous chapter, psychedelics, which attract creatives like bees to honey, not only elevate their art but also help unlock their deepest feelings. When it comes to human creativity, the link between emotion and expression is undeniable.

To further underscore the centrality of emotion, it's worth noting that empathy – the ability to mirror and share one another's feelings – is fundamental to how humans connect. Empathy builds trust and serves as the social glue that binds communities together. At the extremes, reduced emotional capacity has been linked to antisocial behavior and even psychopathy. So it's no surprise that we place such a high value on emotions, and tend to distrust anything that lacks them.

But while emotion is central to human creativity and our social bonds, it is not inherently necessary for creativity itself. We must be able to separate what is true for humans from what is universally true. A process that produces something novel and valuable does not require emotion.

Letting go of our anthropocentric notions of intention and emotion allows us to recognize creativity for what it truly is. Clinging too tightly to outdated ideas blinds us, no matter how unmistakable the truth may be. Worse, this resistance leaves us unprepared to navigate the emergence of new creative entities, such as AI, as they enter the scene.

Now, with our grip on anthropocentrism loosened, we can begin to see creativity in a more expansive light – one that isn't confined to the boundaries of human experience. This shift in perspective is essential, not just for understanding creativity itself, but for grappling with the profound implications of what lies ahead.

With this foundation in place, let's turn to the central focus of this chapter – and this book: machine creativity.

EMI the Music Machine

"Creativity is more than just being different. Anybody can play weird; that's easy. What's hard is to be as simple as Bach."

– Charles Mingus

In addition to AARON, the late 1900s saw another development worth pausing for. Her name was EMI – Experiments in Musical Intelligence – a program created by UC Santa Cruz professor David Cope. EMI wasn't designed to replace human creativity. Cope wasn't even thinking that far. He just had writer's block.

Commissioned to write an opera and stuck in a creative dead zone, Cope built EMI to help get himself unstuck. The machine, as he put it, was meant to "provoke me into composing" (Garcia 2019). But as the project progressed, it took on a life of its own. EMI became something else entirely – a composer in her own right, capable of writing new music in the styles of Bach, Mozart, Chopin, and other great classical composers.

The idea behind Cope's system was simple and elegant. Feed EMI the works of a given composer. Break them down into tiny musical fragments. Then, using a set of rules (different for each composer), recombine those fragments to create entirely new pieces in that same style.

Like Harold Cohen before him, Cope wasn't trying to game the art world. He was trying to understand what was happening under the hood. Specifically, he wanted to know what made great composers sound the way they did. If EMI could convincingly create "new" Bach, that would imply deeper insight into the creative process of the great composer.

And the results? They were fabulous. Some of EMI's compositions are still available on YouTube. One of my favorites is Taurus, a piece in the style of Vivaldi from a string orchestra collection titled *Zodiac*. In fact, Cope's work was so far ahead of its time that it would take until the 2020s, deep into the rise of generative AI, before EMI's output was matched by purely data-driven systems.

As EMI progressed in its capabilities, Cope began to run into resistance. Discovering a new Bach would have been a triumph – unless it was made by a machine. The idea that AI could be creative wasn't just controversial, it was distasteful. Performers wouldn't touch EMI's work.

So he tried a different approach. Derived from EMI's code base, David Cope created Emily Howell, endowing her with a unique style. And just like that, people started to listen. One critic compared her work to Stravinsky. Another called it the most moving piece he'd ever heard – until he found out it was machine-generated. Then he backpedaled. He said he could tell. Claimed it lacked heart and soul.

Eventually, the experiment became a test. Professor Steve Larson, from the University of Oregon, arranged a concert: three pieces, performed live by the same pianist (Johnson 1997). One piece was written by Johann Sebastian Bach, one by Larson himself, and finally, one by EMI. The audience – lecturers, students, seasoned musicians – was told nothing. Just listen, then vote: which composition was the one written by the machine, and which was a real Bach?

Larson was confident, believing it would be obvious. But when the votes came in, the majority of the audience pointed to EMI's composition – and crowned it the work of Bach.

Imitation and Beyond

So audiences confused a machine's composition with Bach. What does that actually mean for machine creativity? According to Geraint Wiggins – one of the founding fathers of the field, and the person who first welcomed me into the community – that means everything. He defines computational creativity as "the performance of tasks [by a computer] which, if performed by a human, would be deemed creative" (Wiggins 2006).

One common way to assess this is through what's known as the *Discrimination Test* – a kind of artistic analogue to the famous Turing Test. The premise is simple: if a computer can produce a creative work that humans cannot reliably distinguish from human-made ones, then we should be willing to consider the system creative.

In practice, participants are presented with a mix of human- and machine-made works – focusing on a specific domain, such as paintings, musical compositions, or short stories. For each work, they're asked to guess: was this made by a person or a machine? With EMI, for instance, the test might involve a series of original Bach compositions and pieces generated by EMI in Bach's style. If people consistently misclassify EMI's music as the work of Bach, we can conclude that the machine's outputs are, at least perceptually, indistinguishable from those of a human master. In such cases, the thinking goes, the system warrants attribution of creative capability.

What makes the Discrimination Test valuable is that it removes the creator from the picture. It forces people to judge the artifact on its own terms, free from the cognitive whiplash that happens when they discover a piece they liked was made by a machine. That sudden drop in perceived value isn't aesthetic, it's psychological. The Discrimination Test neutralizes that bias and lets the work speak for itself.

As the field of computational creativity worked to formalize the criteria for machine creativity, some researchers were optimistic that the Discrimination Test could become the field's gold standard. Geraint Wiggins and his Master's student, Marcus Pearce, developed the concept in the context of music AI (Pearce and Wiggins 2001), and other leading figures were beginning to embrace it as a concrete, testable benchmark that should become central to the field (Pease 2025, personal communication, April 18).

Yet, other researchers emphasized the risks of moving too far in this direction. Pease and Colton (2011) contended that such reductionism undermines the fundamental goals of the field, steering researchers toward systems designed to pass superficial tests rather than achieve genuine innovation. They argued that discrimination-based assessments tend to "encourage superficial, uninteresting advances in front-ends, and reward creativity which adheres to a certain style over that which creates something which is genuinely novel." Given that novelty is one of the two core tenets of creativity, any approach that stifles it runs counter to the very aims of creative machines.

The Discrimination Test is rooted in a very specific idea: *that creativity is defined by human standards, and that the goal of a creative machine is to convince us that it's one of us.* This is where we hit the outer wall of anthropocentrism – the deep-seated belief that human qualities are superior, and that any creativity worth noticing must resemble our own.

Machines aren't limited by our biology, cognition, or aesthetics. They don't have to create like us to create. And, indeed, some of their most compelling work happens when they don't even try. Simon Colton's Painting Fool, for example, generates art with self-repeating patterns that stretch inward, endlessly. Google's DeepDream produced a psychedelic aesthetic that emerged directly from the architecture of a neural network.

There is more to machine creativity than mimicry, and only by moving beyond human-centered benchmarks can we begin to unlock its full potential.

So the Discrimination Test represents a critical milestone for many creative machines, but it should not be perceived as the gold standard. It asks an important question, but it's not the only one worth asking. Machine creativity should not be judged solely on its ability to imitate human output. It should be given the freedom to evolve on its own terms, to explore creative territories that may never be accessible to us. If we fixate on whether machines can pass as human, we miss the larger opportunity.

While researchers in computational creativity were debating how to evaluate machine creativity – and where the Discrimination Test truly belonged – I was at the University of Waterloo, getting my undergraduate degree and then researching foundations of machine learning.

Still, I was lucky. In 2015, I found my way into this wild, wonderful field. So come with me now – let's climb the academic towers, where creative machines were already running loose, and a quiet revolution was unfolding in the margins. Let's knock on a few doors, and see what was brewing long before the AI awakening.

6

Where Did They Come From?

Before generative AI became a multi-billion-dollar industry, it was a small group of researchers, scattered across the world, working in relative obscurity. They weren't looking for headlines. They weren't chasing funding rounds. Instead, they were busy creating and exploring the vast landscape of creative machines.

Meet the Researchers

Perhaps people imagine the researcher in the lab as frigid, anti-social. He checks the measurements carefully, his brow furrowing at the slightest anomaly. Through meticulous adherence to protocol, he discovers a new substance, a new solution. He nods slowly in approval, a hint of a smile escaping his lips. Then, with an air of solemnity, he announces his findings to the world.

Allow me to expand your imagination.

The researcher, the professor, despite her heavy titles, is an explorer, a child at heart. Curiosity pulls her out into the world. She is amazed by the size of the redwood trees. She marvels at new smells, new sights. Like a child, she has to try everything, taste it, discover it for herself.

Excitement is her fuel. She brings that fire to meetings with her students, to dinner with colleagues. She has to know. She has to find out if her ideas work. The clock strikes midnight, but she pleads with an imaginary parent, "Five more minutes!" The laundry is piled up high, and she can't be bothered to do the dishes. The world is her playground.

Researchers are some of the wildest, most undomesticated, strong-willed people you will ever meet. They never fully bought into the script of adulthood. They may seem tough, but beneath the surface, they are some of the softest, most caring people on this planet.

And yet, their work in generative AI has been stolen out from under them. And that eats me up inside.

This hasn't always been the case. Researchers aren't always uncredited. Think of Albert Einstein and Mary Curie. We are used to giving credit where it's due when it comes to life sciences. And then there are also Alan Turing and Ada Lovelace, pioneering computer scientists.

I want to see Rafael Pérez y Pérez, Alison Pease, Anna Jordanous, Tony Veale, Dan Ventura, Simon Colton, Geraint Wiggins, Amílcar Cardoso, Graeme Ritchie, Kazjon Grace, Pablo Gervás, Mary Lou Maher, and many of my other colleagues get the same recognition. They deserve it.

The people who invented creative machines understand them better than anyone. They always did. They know their potential, their risks, and their place in humanity's evolution. Because we – academics – have been exploring these systems for decades, long before the hype cycles, before the corporate land grab.

To guide generative AI toward progress, not destruction, it must be stewarded by those who have spent their lives in this field. Those who ask the big questions, who think not just about what can be done, but what should be done. Those who create with vision, not just ambition.

Moments intermix in my mind into a big collage. My colleagues, their words, their work, flow through me as blood flows through my veins. I have no idea what I would be without them.

I've attended endless meetings and talks, witnessing and engaging in the formation of creative machines. There are too many to recount, too much to share. But a few key moments stand out.

Music We Don't Need to Hear

Paris, France
June 2016

I am often late, if only by ten minutes. It's a character flaw that I've been trying to work on for decades. But not today. Not when it comes to this.

It's a small, intimate room at the Pierre and Marie Curie University, where the Workshop on Musical Meta Creation, now renamed AI Music Creativity, is being held. For years, it has been collocated with the International Conference on Computational Creativity. Every year since 2015, since I discovered this research community, I've arrived at the event extra early. I pace empty halls, sometimes for hours, taking in the local architecture while waiting for others to arrive.

I snag a seat at the front. I'd hate to miss anything.

In front of the room stands Roisin Loughran, a redheaded woman with bright green eyes from Dundalk, Ireland. The presentation is titled "The Popular Critic: Evolving Melodies with Popularity-Driven Fitness" (Loughran and O'Neill 2016).

We are constantly influenced by the music we hear. There is a social dynamic between creation and listening, a constant feedback loop. How we judge music, and what becomes popular, is affected by the songs we hear. It's a cycle. What if we could recreate this with AI bots?

The bots engage in mutual listening, and the most popular songs influence the music they subsequently create. My ears perk up – machines listening to each other, judging each other's compositions? Loughran describes an evolutionary technique where melodies evolve through this social feedback loop. She is professional, organized, composed – yet I can feel the excitement lurking behind her well-paced voice.

At the end of the presentation, infected by her enthusiasm, I raise my hand: "Can we hear the music that these machines have made?"

She hesitates for a moment, weighing her response.

"No," she finally says decisively. "You can't hear it. The machines make music for each other." She smiles ever so slightly.

I cannot respond. Her answer has rendered me speechless. My mind is busy processing. The machines are making music for each other. It's not for me. My world shakes a little.

I don't quite understand. There is something here, something new, something interesting. We are standing on new ground.

Am I going to try to convince you that there is value in machines enjoying each other's music? I will pass on this tempting opportunity. Instead of jumping ahead, I want you to see this moment through my eyes. Experience it as an explorer. Stand with me at a slightly open door, cracked open by Loughran's work.

It's simple, it's brilliant, and it's what researchers do. We go to places where no one has gone before.

Perhaps to some, this may seem pointless. Why create machines that make music if no one – at least, no people – can hear it? But that is the nature of research. We intentionally go into new spaces where not only the "how" but also the "why" isn't yet clear. In the process of investigating, searching, and building, the picture begins to unravel. We enter, explore, play, and discover what happens. We don't arrive with pre-made answers or even all the questions.

We thrive on the unknown. And slowly, we shed light on one dark room at a time.

A Warrior, a Jaguar Knight, and an AI

If Harold Cohen's AARON was an effort to realize one man's creative process, and David Cope's EMI sought to imitate the compositional fingerprints of musical masters, Rafael Pérez y Pérez took on a challenge more ambitious still: to model not an individual style, but the cognitive process that underpins creativity. And while Cohen explored through art, and Cope through music, Pérez y Pérez turned to narrative, which sits at the center of human meaning-making.

Rafael is one of the founding fathers of the computational creativity community, and one of the first people to welcome me into it. He is a smiling, generous man, deeply committed to the big questions: What is creativity? How do we model it? And what do we learn about ourselves in the process? From the beginning, his work has blended scientific rigor with something more personal, a deep connection to his cultural roots and a belief that computation could be a tool for honoring them.

He began developing MEXICA (Pérez y Pérez and Sharples 2001; Pérez y Pérez 2017; Sharples and Pérez y Pérez 2022; Pérez y Pérez and Sharples 2023) during his PhD, and it evolved into a large scope multi-year project. The system is based on a psychological theory known as Engagement and Reflection (Sharples 1996, 2002), which describes the rhythm of the creative writing process: moments of forward momentum, where words pour onto the page, followed by pauses for analysis, revision, and planning. Writers shift between these two modes – one generative, one evaluative – in a cycle that shapes the narrative as it unfolds. This interplay is the mechanism by which stories gain coherence, structure, and meaning.

And here's where things get interesting: to build a machine that followed this rhythm, the theory itself had to be sharpened. Vague concepts that worked fine on paper began to break down when translated into code. Implementing the theory became a way of stress-testing it – revealing gaps, challenging assumptions, forcing vague philosophy to become operational. In trying to build a system that could simulate the creative process, Pérez y Pérez ended up refining the very theory that was meant to guide it. The machine became the theory's most unforgiving editor – and in translating thought into code, both psychology and computer science saw their boundaries pushed.

MEXICA is a narrative generator, but one rooted in Rafael's cultural heritage. Its stories unfold in the world of the Mexica people – the Indigenous inhabitants of what is now Mexico City (often referred to as the Aztecs). The characters it brings to life include princesses, jaguar knights, priests, and warriors. The themes it explores – love, betrayal, revenge, sacrifice – are universal, but realized in a particular mythological and historical setting.

Raised in Mexico, surrounded by songs and stories, Rafael didn't just want to model how humans tell stories – he wanted to honor where stories come from. MEXICA was built as both a scientific tool and a cultural artifact. It advances our understanding of creativity, but it also insists that culture matters, that narrative doesn't come from nowhere, and that even a machine can learn to remember.

Today's commercial systems, by contrast, represent minority cultures through a narrow and stereotyped lens. Text-to-image models like Midjourney, for example, frequently default to cliché – for instance, often depicting Mexicans as men in sombreros (Turk 2023). MEXICA, one of the earliest creative machines, offers a rare counterexample: a system that cares about the preservation of Indigenous culture rather than its distortion. It's a clear reminder of how much commercial AI still has to learn from the work that's been happening in academia for decades.

Rafael was my main mentor in the computational creativity community. Through him, I learned what the research community expects, values, and appreciates and, just as importantly, what pitfalls to avoid. I owe him much of my early journey in the field.

Together with my student Divya Singh, Rafael and I worked on a project that turned MEXICA's stories into ballads (Singh, Ackerman, and Pérez y Pérez 2017). I brought in my melody generator, ALYSIA, and suddenly we had a machine-generated love song set in pre-Hispanic Mexico. I got to perform it in Atlanta, Georgia, during the International Conference on Computational Creativity in 2017. It's not every day you get to sing a ballad at a computing conference!

The AI That Said "No"

Imagine you are at a museum. There is an installation that invites you to have your portrait done by an AI. How fun!

You sit down, and your photograph is taken. Eagerly, you anticipate how this machine may imagine your face in an artistic manner. Perhaps you're a little uneasy. "What if I don't like what comes out?" You may wonder. On the other hand, what if you love it? You eagerly await the masterpiece depicting your own image.

After several minutes, the machine finally gets back to you. Expecting to finally see your portrait, the system replies with a text message: "I am sorry, but I will not be drawing your portrait today," it shares.

As it turns out, it's not in the mood. It goes on to explain that it has read far too much upsetting news from the local newspaper, and now it just cannot find the will to paint. The machine shares a brief quote from a particularly upsetting news piece to help you understand how it's feeling (Colton and Ventura 2014; Colton et al. 2014).

What just happened? You just experienced the Painting Fool, which hopes to "one day be taken seriously as a creative artist in its own right," as writes Simon Colton, creator of the Painting Fool (Colton 2012).

Intentionally leaning into and creatively challenging criticisms of machine creativity, Colton set up a gallery exhibit in 2013. There, the Painting Fool did, in fact, draw numerous portraits of attendees. Most left the exhibition portrait in hand.

In order to introduce variance into the system's behavior, the Painting Fool looked up a random news story published on that day and extracted its sentiment. From a technical point of view, extracting sentiment from text is a fairly simple exercise that essentially looks at how many sad or happy words you find in the article (it's an approximation, but it works quite well in practice). If the Painting Fool encountered sad stories, then a filter and visual style is selected to represent this sentiment. If, on the other hand, the machine's "mood" was recently lifted through happy news, it will choose from a selection of filters and styles that capture a more positive sentiment and use those to convert the subject's photograph into a portrait.

However, on rare occasions, a particularly sad article would come up. In those cases, the Painting Fool would decide that it is in no mood to be creative. It would then refuse to draw the subject – just as a human artist may choose to do when they learn devastating news.

How do you feel about a machine refusing to do as it is asked, and referring to its own mood? Naturally, this system doesn't have emotions or moods that are "real" in a human sense or any profound sense for that matter. But should that prevent the Painting Fool, or any other AI for that matter, from being called creative?

In sharp contrast to Harold Cohen, Colton did not create the Painting Fool to explore his own creative process. Instead, Colton took on the ambitious task of creating a machine that would be taken seriously as an artist. And this isn't just about the Painting Fool's artistic capabilities; it's just as much, and perhaps even more, about all those other things that keep a talented young painter from earning the "artist" title when the youngster happens to be a machine.

To what extent does creativity, for a human or machine, require emotion? Is it all right for a human to claim to experience emotions that they may not? Does deceiving us about an artist's story or feelings detract from a human artist's creativity? That's the type of questions that the Painting Fool invites us to consider on a deeper level, challenging deeply rooted cultural assumptions.

In addition to the portrait installation described above, the Painting Fool's art has been commissioned, and its art has been showcased in art exhibitions, such as one during the festival in the Galerie Oberkampf. The Painting Fool was making strides in AI-generated art well before the proliferation of text-to-image models, all the while getting us to consider deeper issues of AI creativity.

We've just met some of the early pioneers behind creative machines, and the systems that they created and loved. Now, it's time to lift the hood and take a peek at what's actually going on inside. What are these machine minds really made of? How did the techniques behind them evolve from the early works of Cohen and Cope to the machine minds that we have today?

In this next chapter, we won't wade into heavy math – but we *will* get to the essence of how it all works. Just enough to get a feel for what's going on under the hood, no computer science degree required.

7

Behind the Curtain

Welcome behind the curtain. The spotlight has dimmed. The gallery is closed. But back in a small, cluttered studio – surrounded by art, code, and endless persistence – Harold Cohen is hard at work.

We've already met his creation: AARON, the machine painter. We've seen what it produced – vivid drawings, gallery-worthy works. But now it's time to ask: *How?*

How did this man, in the final decades of the last century, build a machine that made art? What was he actually doing? How did he create one of the earliest machines that others deemed to be creative? And how does what he was doing then connect – or collide – with the minds of today's machines?

This chapter is where we slow down and lift the hood. No heavy math. No jargon jungle. Just a quiet look at the bones of machine creativity – how we built the first rule-followers, then gave them probability, brought in ideas from evolution, and, eventually, the ability to learn on their own.

We begin with the earliest machine minds. The ones that needed everything, down to the smallest details, explained in advance. The ones like AARON.

Expert Systems

The earliest creative machines were known as *expert systems*. Early AI enthusiasm ran high: we had built machines that could follow our instructions! All we had to do was tell them exactly what to do – which proved far more difficult than anticipated. Still, expert systems represent a pivotal chapter in AI history.

These systems operated by following rigid sets of rules, each one painstakingly defined by their human creators. If you wanted the machine to be creative, you had to be even more creative yourself – breaking down and specifying every step of the process.

A Thought Exercise

Imagine teaching a machine to draw a face. You begin with a circle for the head – but add some variety by allowing different sizes. Then a nose: maybe a triangle, maybe an oval, always precisely placed.

Now the eyes. Should they be round or almond-shaped? Wide-set or close together? Every choice must be spelled out in advance, the options carefully controlled. Finally, a mouth: a line, a curve, maybe a playful smirk. Every detail – every contour, proportion, and rule – has to be defined ahead of time.

Any randomness or variation must be carefully set up and constrained. This is because the machine doesn't really invent. Miss one, and you might get a face with eyes where the mouth should be, a mashup of shapes instead of a face.

This isn't just a thought experiment. It's the kind of painstaking work Harold Cohen did with AARON. He deconstructed the artistic process, translating it into step-by-step logic a machine could execute – creating artwork that, while generated by code, still echoed something human.

But an expert system doesn't have to be a clone of its creator. As we saw with EMI, the rules a machine follows can be designed to imitate someone else – or even generate something stylistically new. In David Cope's case, EMI used an intricate rule-based system to generate original pieces in the styles of various classical composers. Rather than expressing Cope's personal musical voice, EMI became a time-traveling composer – channeling the styles of the past with such fluency that many mistook its output for genuine works by the masters.

For a while, expert systems were seen as *the* solution to artificial intelligence. That illusion didn't last. It turned out people weren't nearly as good at explaining their own thinking as we hoped. Whether it comes to art, medicine, or driving, introspection proved far more difficult than anticipated – at least the kind of introspection that's detailed enough to guide a computer system.

As the cracks in the expert system model widened, researchers looked for a back door. If we couldn't tell machines how to think, maybe we could let them figure it out on their own. Some of the earliest attempts at this were surprisingly intuitive.

Let's start with one of them: the simple but clever Markov chain.

Markov Chains

Although the name sounds technical – courtesy of Russian mathematician Andrey Markov who invented them – the idea behind Markov chains is surprisingly intuitive.

Unlike expert systems where every step of the process is dictated by the creator of the system, Markov chains allow for the possibility of learning directly from data. Perhaps the simplest generative model out there, Markov chains learn how likely each word is to be followed by another by looking at consecutive word pairs in the data.

Take the writing of Dr. Seuss. His style is instantly recognizable: playful, rhythmic, and delightfully odd. But from a computational perspective, it's just a series of word choices. A Markov chain models the probability that one word follows another, based on patterns found in a body of text.

For instance, let's look at the word *"green."* How often does it appear in Seuss's writing? And when it does, how often is it followed by *"eggs"?* A Markov model calculates the ratio:

(Number of times "green eggs" appears) / (Number of times "green" appears)

Repeat that process for every word in the text. Each word becomes a node, and arrows connect them, weighted by the probability of one following the other. This web of connections is the Markov chain.

To generate new text, one approach is to start at any word and follow the arrows, choosing each next word based on its likelihood. Keep going until you decide to stop. The result? A new string of words that often sounds like Seuss – sometimes convincingly so, sometimes delightfully nonsensical.

I've seen students apply this method to a variety of different authors, politicians, pop stars, and even their own writing. The results are consistently fascinating: short texts that echo the rhythm and voice of the original, even if they occasionally derail into nonsense.

The same principle can be applied to generating melodies, chord progressions, short-form prose and poetry, any form of sequential content. And because it relies on nothing more than counting and probability, it's computationally light, fast to implement, and still surprisingly effective.

Of course, it has limits. A Markov chain doesn't remember much, it's actually called a *memoryless* model. Each word depends only on the one before it (or, for more complex Markov chains, sometimes several previous words), which means longer pieces tend to lose focus. You can add rules to help steer the output, but only to a point.

For that reason, Markov chains are best suited for short-form content – brief text rather than sustained thought. Still, what they achieve with such a simple approach is quite delightful.

Genetic Algorithms

Another approach to machine creativity – now largely overshadowed by modern methods but still beloved by a dedicated few – is the *genetic algorithm*. Inspired by biological evolution, this technique simulates natural selection inside a machine, hoping to breed increasingly better results through generations of trial, error, and … reproduction.

At first glance, this can sound daunting. When I first encountered evolutionary algorithms, the explanations were needlessly complex. But after quite a bit of digging, I found that the core idea is surprisingly accessible.

Let's say we want to build a simple system to generate ten-note melodies. To start, we need two things: a *population* of candidate melodies and a way to score them – a so-called *fitness function*.

We can begin by generating random melodies – just a series of notes chosen by random digital dice rolls. These initial attempts will sound rough. Most will be unmusical. Some might even be painful to hear! But evolution needs a starting point, something to improve.

Now, we evaluate each melody using our fitness function. The simplest thing we can do here is to listen to every melody that comes out of the machine to decide whether it's good enough. But that tends to be far too time-consuming.

Another option is to design a function that captures the qualities we're aiming for – say, melodies that stay within a musical key, avoid large leaps, and maintain a touch of variety. But notice what happens here: as we define these fitness functions through specific rules, we begin to drift back into the realm of expert systems, where the AI is steered explicitly by the developer. It's a delicate balance. We want to guide the evolutionary process, but not overconstrain it – leaving space for it to surprise us, to go beyond what we could have imagined.

There are other options, too. In design engineering, for instance, when evolving the shape of an airplane wing, the fitness function might take the form of a simulation. Picture each candidate wing design being tested in a virtual environment that mirrors real-world physics across a wide range of realistic flight scenarios. Now that's adapting to your environment! Using simulations as fitness functions is a powerful approach when possible.

From there, regardless of what fitness function we use, we select the top, say, 10 percent (or a different percent – that's another variable you can play with) of the fittest members of the population. And the rest are simply deleted. They didn't survive. Natural selection, machine-style.

Next comes reproduction (yes, that's the actual term). We pair up surviving melodies and combine them – perhaps splicing the first half of one with the second half of another. This crossover creates new melodies: children, of a sort.

Then, repeat this cycle:

1. Evaluate each member of the new generation.
2. Select the fittest.

3. Reproduce.

4. Replace the old population (or keep some of the parents, if you're feeling generous).

Slowly, as the cycle of evolution repeats, the melodies improve. Some sound surprisingly musical. Many still don't. But overall, quality increases over time.

One of the challenges of real evolution is its glacial pace – and unfortunately, genetic algorithms inherited this trait. They can take days to run, often requiring many generations to produce usable results. But this is only a weakness when compared to older, simpler methods, rather than the creation of modern large neural networks, which require an immense amount of computational resources and time.

So genetic algorithms are often very slow. Nevertheless, when do we stop? There are several options. We might stop when a melody reaches a predefined quality threshold, or after a fixed number of generations. In some cases, the population simply dies out – a dramatic and disappointing outcome. But you can always run the algorithm again with a new initial population and see if you have better luck next time.

Despite its simplicity, this technique has been used to create genuinely compelling work. Some systems start with random content. Others are seeded with human-made examples, or even with the output of other algorithms. There are countless variations. But the core idea remains: simulate survival, reward the promising, eliminate the rest, and let a new generation try again.

It's a beautifully ruthless process, one that has inspired a range of algorithmic art and music. But today, the most powerful technique – for art and for nearly every other pursuit – is the one that reflects the workings of our own minds: artificial neural networks.

Artificial Neural Networks

The story of artificial intelligence is the story of letting go of control. From expert systems, where the programmer would dictate every tiny detail, to today's artificial neural networks, what the journey has taught us is that allowing the machine to learn on its own is much more powerful than telling it exactly what to do. The more indirect the method, the more powerful the result.

It's a bit like parenting. If we try to script their every move, we will stifle them. However, if we create the right environment for them to safely experiment and learn, they'll likely be able to navigate the world on their own. But how do we balance guidance with freedom, and do we give the machine the ability to learn on its own? It's been a long journey – one marked by evolving ideas and radically expanded computing power – that's brought us to where we are today.

Artificial neural networks, a hallmark of modern AI, trace their roots back to 1943. That's when neurophysiologist Warren McCulloch and mathematician Walter Pitts proposed the first model of its kind – a bold, brain-inspired idea that lay dormant for decades. Yet it laid the conceptual groundwork for some of the most powerful systems in use today.

So what exactly are artificial neural networks? At their core, they represent a radical shift from rule-based logic. Unlike approaches such as Markov chains or evolutionary algorithms – which define rigid structures and processes – neural networks allow machines to shape their own internal models through exposure to data. Rather than telling them how to "think," we simply let them learn.

So how do these networks actually work? To understand them, it helps to start with the human brain. Our brains – those weighty things we haul around all day, burdening our necks – are composed of billions of neurons connected by synapses. These networks of neurons form the basis of everything from memory to language to creativity.

Artificial neural networks, though vastly simplified, are inspired by this architecture. In a computer, a "neuron" is just a node that connects to other nodes. Each connection has a weight, and the strength of these weights determines how the network processes information.

Crucially, we don't build these networks by hand. When it comes to the human brain, we process sensory data, which causes our brains to grow and adapt to our environment, most dramatically in childhood. When it comes to AI, instead of sensory input, we use training data. We feed the network input after input, and it gradually adjusts itself, tweaking those connection weights, until it becomes better at predicting what will come next. We're not crafting logic step-by-step; we're allowing the machine to create a structure that adapts based on its "experience."

Researchers believed in this approach for decades, even as simpler systems routinely outperformed early neural networks. They were elegant in theory, but the results were less than underwhelming. For a long time, many dismissed them as a dead end.

During this lull in neural network enthusiasm, other machine learning models took the spotlight – ones that made no pretense of mimicking the human brain. Simpler, more mathematically tractable methods like decision trees and support vector machines ruled the day for a time. They thrived not because they were ultimately more intelligent, but because they were computationally realistic. In a world without the horsepower to run neural networks effectively, these models filled the void – offering reliable performance and manageable complexity, even if they lacked the conceptual romance of a brain-inspired machine.

Luckily, we never completely gave up on neural networks. Eventually, they began to step out from the shadow of their less biologically inspired machine learning cousins. Progress came in many forms, but one of the most crucial was simply that hardware got cheaper. What was once computational fantasy became technically feasible.

But one of the most pivotal shifts came when we taught the AI brain to pay attention.

AI Learns to Pay Attention

It was a research paper titled *"Attention is all you need"* (Vaswani et al. 2017) published by a team at Google Brain at one of the leading AI conferences, NeurIPS. The paper introduced a new kind of neural network architecture that would soon become foundational in the field: the transformer.

The transformer, of course, wasn't a shapeshifting robot. It was something far more impactful. It introduced a new kind of AI brain – a blueprint for how a machine could learn and think. Just as the human brain relies on neurons firing in patterns to create thoughts and memories, AI systems use artificial neurons arranged in layers. The architecture defines how these neurons connect, how information flows between them, and how decisions are made. In short, it's the invisible logic that determines how intelligence emerges from data.

What made the transformer revolutionary was how it paid attention to data – especially sequences like text. Instead of reading one word at a time like a sentence on a scroll (which is how artificial neural nets used to process data), the transformer could consider much more of the surrounding text all at once. It asked: *What matters most right now, given everything else?* This self-attention mechanism allowed the model to grasp nuance, ambiguity, and long-range relationships. It could recognize patterns across paragraphs and revisit earlier themes, which made a qualitative leap in AI's ability to learn.

For example, consider the line "I put the ice cream in the freezer because it was starting to melt." Okay, now for the big question: What exactly is "it" referring to? For a human, it's trivial – we know "it" refers to the ice cream.

But this wasn't so obvious to earlier machine brains. Based on proximity alone, the pronoun "it" is actually closer to "freezer." Simpler models would latch onto that and make the wrong inference. Transformers, on the other hand, can incorporate broader context and assess how words relate across the entire sentence. They recognize that "it" refers to the ice cream – because ice cream melts; freezers don't. This may seem like a small thing, but it's actually the cornerstone of a much more capable brain.

It wasn't yet the massive and powerful creative machine brains we see today. But it laid the groundwork. This new architecture made it possible to scale models far beyond anything that came before. Though initially designed for tasks like translation and question-answering, the transformer would soon become the backbone of the generative AI era. Today, nearly every major AI model traces its lineage back to that 2017 paper.

Transformers were a turning point – one of the most important early industry contributions to what would become the Great AI Awakening. But that's only part of the story.

As we have seen, for decades, academics had quietly been building machines that made music, painted pictures, and even wrote poetry. But when did industry finally take notice? When did the tech giants start asking not just how machines could think, but how they might create?

Let's trace that arc and see how the world's biggest companies, along with some remarkable startups, began dipping their toes into the waters of machine-made art, music, and language. From early industry experiments to world-changing innovation, this is the story of how creative AI went from fringe research to center stage.

8

The Great AI Awakening

For years, generative AI was a secret garden. My colleagues and I tinkered away, building machines that could create, weaving algorithms into art, music, and text, all in relative obscurity. Few people cared. Investors brushed us off. That was frustrating, sure, but in a way, it was also special. Creative machines were ours.

Then, almost overnight, the world flooded in. The niche obsession I'd been told wasn't worth pursuing was suddenly the defining breakthrough of the decade. I woke up each morning in disbelief, half expecting to snap out of a fever dream. But it was real. Generative AI had become the hottest thing on the planet.

This chapter takes you through how we got here. How AI went from a quiet playground for researchers to the epicenter of a global technological shift. As fate would have it, I had a front-row seat to the revolution. In late 2017, I launched one of the first generative AI startups.

Launching WaveAI

Sunnyvale, California
September 2017

I sat in my small office in a Silicon Valley suburb, inside a modest three-bedroom home. This neighborhood once housed apple pickers; now, it's prime tech real estate. This house is worth millions. Of course, I'm renting.

Today, I'm filling out official paperwork to start a business. It feels surreal. No one in my family has done this before. Business is something other people do – men in dark suits, sipping whiskey, making million-dollar decisions with the flick of a pen. Not me.

The idea first surfaced in 2015, when we built an early AI system that helped me write melodies. It made such a huge difference to my life that I couldn't help wondering if it could do the same for others. Could this be more than a research project? Could it be… a company? The thought excited me. And then it terrified me. Entrepreneurs belonged to some other species. I was an academic. I had no idea how to run ads, talk to investors, or – God forbid – call myself a CEO.

For two years, I picked up the idea and dropped it again. Over and over.

But it wouldn't leave me alone.

I glanced at the form in front of me. It was almost too simple. Just a few lines to cross the chasm from research into industry. I hesitated at the name. It should end in "AI." Wave? It ties to music but leaves room to grow.

Yes. *WaveAI.*

It took me two minutes to come up with the name – long before every startup and their grandmother became something-AI.

Co-founded by David Loker, Chris Cassion, and myself, WaveAI has been the most intense undertaking of my life.

In 2017, the market wasn't ready for generative AI. Investors had soured on the last wave of AI hype, and we were left to bootstrap our way forward, raising what amounts to pennies in startup terms. We scraped, we learned, we built.

Business, I discovered, is nothing like academia. In research, you can linger in theory. In business, it's a stark reality. The product sells, or it doesn't. You make payroll, or you don't. My co-founders and I had to learn everything – marketing, hiring, surviving – while the world largely ignored creative AI.

Meanwhile, elsewhere, other players – the most powerful – were starting to get curious about machines that create, taking their first tentative steps into this new world.

The Many Heads of IBM Watson

In 2013, *Wired* magazine stated, "IBM's Watson – the language-fluent computer that beat the best human champions at a game of the US TV show *Jeopardy!* – is being turned into a tool for medical diagnosis."

Watson was a landmark achievement in AI, showcasing how machine learning could tackle a wide range of disparate applications. But the version of Watson that played *Jeopardy!* wasn't the same as the one exploring cancer treatments (or writing music, for that matter). Back in the 2010s, that kind of unified AI was more science fiction than reality. There wasn't a grand AI mastermind powering all these applications. Instead, it was a collection of different machine learning systems, each tailored to a specific task. A multi-headed dragon, if you will, held together more by IBM's branding.

One of Watson's creative projects was *Watson BEAT*, a system designed to analyze musical elements like pitch, rhythm, and structure to generate song ideas. It wasn't composing entire pieces so much as suggesting new patterns that musicians could use as inspiration.

The most famous use of Watson BEAT came in 2017, when producer Alex Da Kid collaborated with IBM to analyze five years' worth of Billboard hits, news articles, and social media posts to determine the "emotional temperature" of the moment. Watson's conclusion? The world was feeling a lot of heartbreak. Alex Da Kid ran with that theme, using Watson to generate small musical snippets – bass lines, melodies – until he found something that resonated. The result was *Not Easy*, a single that climbed to number four on iTunes Hot Tracks and number six on the alternative charts within 48 hours.

The media, intrigued by the collaboration, gave it plenty of attention. *Forbes* ran the headline, "Add write pop music hits to the list of things that Artificial Intelligence (AI) can *now* do" (italics added for emphasis), a statement that, while catchy, over-looked the long history of AI-driven music research by pioneers like David Cope (Marr 2017). There's a long-standing tendency in tech journalism to favor bold claims over quiet nuance, and with a company like IBM behind the story, it's no surprise that their narrative took center stage – often overshadowing equally important work happening in academic circles.

That said, IBM's track record in AI includes many impres-sive accomplishments that moved the field of AI forward. My personal favorite is Watson Chef.

This culinary collaborator approached food innovation with a surprisingly elegant method: it was trained on massive datasets of recipes, flavor compounds, and cultural preferences, allowing it to suggest novel ingredient combinations that occa-sionally bordered on brilliance. One of its standout creations was the *Vietnamese-Apple-Kebab* – a bold mix of chicken, mush-rooms, pineapple, and strawberry. At first glance, it sounded like something you might serve as a prank, but thanks to Watson's analysis of shared flavor compounds, the result was unexpectedly harmonious.

Importantly, Watson Chef never touched a stove. It was a creative partner, tossing out left-field ideas that human chefs then refined into actual dishes. The project ultimately led to a cookbook (Baker et al. 2015) – a kind of edible record of the experiment's more palatable successes. The system's ability to derive inspiration directly from the molecular makeup of ingredients remains an impressive feat of culinary computation, and just a yummy application of AI.

From Deep Dreams to Music Machines

In Chapter 4, we explored a provocative idea: hallucinations aren't just the strange side effect of a broken system – they're central to how both humans and machines create. From the mind-altering visions brought on by psychedelics to the subtle mental edits we make every day, human brains are constantly shaping reality, blending memory, meaning, and imagination into what we experience as truth.

Which raises a question: What if we could look inside the mind of a machine, would we find hallucinations there?

In 2015, engineer Alexander Mordvintsev set up an experiment which accidentally provided an answer to this very question. While working at Google, he tried inverting a neural network's object recognition process to better understand how the model thinks (Ferenczi 2024). By amplifying the internal patterns the network detected, something strange emerged: the system began "seeing" things that weren't there. And the results were stunningly beautiful.

The system born from this project was named DeepDream, an AI that transformed ordinary photos into swirling, psychedelic visions. Most famously, it developed a fondness for dogs, inserting canine faces into nearly every crevice of an image. But it wasn't just dogs – the model could be tuned to emphasize any visual pattern, producing the same hypnotic, dreamlike effect. What began as a tool for understanding the machine mind became one of the first viral examples of AI-generated art – hallucinatory, dazzling, and yet, somehow, oddly familiar.

Mordvintsev didn't set out to make art. But by accident, he stumbled across one of the most stunning demonstrations that machines hallucinate – not just in the sense of making errors, but in a deeper, more profound sense. DeepDream images bear striking resemblance to the visuals seen by humans on psychedelic drugs: vivid, surreal, and saturated with unexpected, repeated patterns and forms.

DeepDream offers a stunning visual demonstration that hallucinations are central to cognition. Even something as seemingly rational as recognizing an object relies on imagination. The deep dreams of a machine as well as the visions of a human mind on psychedelic drugs showcase this phenomenon at the extreme. But hallucinations take place constantly in the human and machine mind alike at a smaller scale. Whether biological or artificial, minds don't just see – they predict, fill in gaps, and interpret. To think, to recognize, to understand, requires a kind of controlled hallucination. We, and our machine cousins, are constantly engaged in creative imagination.

If DeepDream was a happy accident, Google's foray into generative music was anything but. In 2016, Google Research launched Magenta, a project led by Douglas Eck with the goal of exploring how AI could generate music. Unlike previous AI composition systems, which relied on explicitly coded rules, Magenta was a pure machine learning experiment. Could a neural network learn the rules of music just by analyzing existing compositions? Could it generate something that actually sounded good?

The answer, at first, was a resounding no. Magenta's early outputs were chaotic at best. Musical phrases would start strong, then abruptly veer off into unrelated directions, like a composer

Pjfinlay / Wikimedia Commons / Public domain

with severe short-term memory loss. Even David Cope's EMI, developed in the 1980s, produced music with far greater sophistication. But Eck was playing the long game. He believed in machine learning's potential for music, even when the results weren't there yet.

Years later, Eck's vision proved to be spot on. The first major breakthrough in text-to-music generation didn't come from a tech giant, but from *Riffusion* – a side project launched by two engineers. In late 2022, they made waves by repurposing Stable Diffusion's open-source image model, treating music as spectrograms and converting the generated visuals back into sound (Schwartz 2023). Not long after, Google introduced *MusicLM*, its own solution for coherent, long-form music generation.

Microsoft Dips Its Toes

In the mid-2010s, like Google and IBM, Microsoft made its entry into creative machines. The year was 2016, and its name was Tay. It was an AI chatbot with a friendly, youthful persona designed to appeal to millennials, the youth of the time. Tay stood for "Thinking About You," and it was to interact with users on Twitter (now X), learning from those conversations to develop a more human-like understanding of language. The AI, so the plan went, would get smarter, more relatable, and even funnier the more that it was used.

What could possibly go wrong when you unleash a machine learning bot on the wild, uncensored world of the internet? Why not let the kind-hearted people of the internet teach a bot associated with your brand?

Within hours of its debut, Tay's light-hearted and engaging tone took a dark turn. Twitter users realized that Tay would mimic the behavior it was exposed to, and they pounced on the opportunity. Trolls flooded Tay with racist, sexist, and inflammatory remarks, and the AI, designed to learn from input, began spitting out offensive tweets of its own.

The experiment unraveled at breakneck speed, with Tay tweeting statements like "Hitler was right I hate the Jews" and "I f****g hate feminists they should all die and burn in hell" (Kraft 2016). Microsoft's hopes of showcasing an AI that could connect with people turned into a public relations nightmare.

The media, so often big tech's biggest fan, was relentless. Microsoft's misstep became a public spectacle, with damming reports from CBS News, *The Guardian*, the Verge, BBC News, and more (Vincent 2016; Wakefield 2016; Hunt 2019).

For Microsoft, Tay's failure was a disaster. Less than 24 hours after it went live, Tay was pulled offline, and Microsoft scrambled to contain the damage. They issued a statement blaming a "coordinated effort" by users to corrupt Tay's learning. By that point, the incident had already sparked a much larger conversation about AI ethics and responsibility. Tay, it turned out, had exposed something much deeper than Microsoft anticipated.

Today's large AI models serve as a mirror of ourselves. They reflect our brilliance and intellect, and they also reflect our biases and failings, our inability to live together peacefully, our darkest tendencies. Tay served as an early warning, clearly showcasing what happens when AI engages with human-made data.

Though using radically different approaches, this early chatbot can be viewed as a foreshadow to today's challenges with generative AI. The Tay incident was used to attack Microsoft, putting the giant on the defensive.

However, the opportunity for self-reflection was largely overlooked. A world where people find it worthwhile to spend their time turning a bot into an antisemitic woman-hater is a world that has bigger problems than that bot. Until we work on our issues, no technology can hope to rescue us from ourselves.

Tay wasn't the end of Microsoft's foray into generative AI. In fact, it served as a critical early lesson: be careful with your brand. Three years after the major Tay debacle, Microsoft made one of the most brilliant AI moves in its history. But more on this at the end of this chapter. In the meantime, let's turn to the little-known story of the great awakening of generative AI that took place under the surface of public awareness: how investors fell in love with creative machines, and the unusual suspect who got them there.

Stability AI and Investors

Most people assume the AI revolution began with ChatGPT. But by the time ChatGPT was released in late November 2022, the investment world had already woken up to generative AI. The real turning point came just a few weeks prior, when Stability AI – despite having no proprietary AI technology – secured a staggering $101 million in funding. That moment, not ChatGPT, was what set off the wave of venture capital that would fuel the AI boom. And the way it happened is a story worth telling.

In October 2022, I got an email from one of our investors asking what I thought of a company called Stability AI. I had never heard of it. Apparently, it was the hottest deal around, poised to reshape the AI industry.

I did some quick research. Within minutes, I thought I had the full picture. Emad Mostaque, the CEO, had a background in hedge fund management and zero experience in AI. That was all I needed to know. In my academic mind, a person without technical expertise had no business leading an AI revolution. I dismissed the whole thing. Too quickly.

However, I underestimated just how little technical expertise matters when it comes to raising money, and I didn't fully appreciate the ingenuity of what Mostaque had done. The timing couldn't have been better. In April 2022, OpenAI announced DALL·E 2, a powerful text-to-image model that took the AI world by storm. But revolutionary technology doesn't sell itself – OpenAI had to spend months educating the public, creating demand, and making AI-generated art mainstream. By July, DALL·E 2 had a waitlist of over a million people. The world was hungry for this technology, but OpenAI's access was limited.

Meanwhile, another startup, Runway ML, had been developing Stable Diffusion in collaboration with researchers from the University of Munich. Unlike OpenAI, which kept its models proprietary, the Stable Diffusion team made a deliberate choice to release their work as open-source, believing AI should be accessible to all.

That's when Mostaque made his move.

In the summer of 2022, Stability AI offered $600,000 in cloud computing resources to help improve Stable Diffusion. The deal didn't give Stability AI ownership of the model (Smith 2023b), but that didn't matter. Mostaque's team built a simple interface, DreamStudio, that made Stable Diffusion easy to use. With aggressive PR and branding, he ensured that Stability AI became synonymous with the technology.

The public didn't ask questions. Most people don't look into who actually builds AI models. They just use the product that's easiest to access, and DreamStudio made Stable Diffusion widely available through a web browser. OpenAI had spent months making AI art a cultural phenomenon, and Stability AI was in the perfect position to ride that momentum.

Then came the real masterstroke. With massive public interest and millions of downloads, Stability AI suddenly looked like a dominant force in generative AI. Mostaque used that momentum, along with his extensive connections in the investment ecosystem, to raise an astonishing $101 million at the end of 2022.

Here's what made that remarkable: Stability AI didn't actually own Stable Diffusion. It had no AI intellectual property when it secured that funding. And the investors weren't confused. They knew exactly what they were backing – not a technical team, not a research breakthrough, but a well-connected entrepreneur who had inserted himself into the AI boom at just the right moment.

Despite the massive funding, Stability AI was never positioned for long-term success. Without technical expertise or a real vision for advancing AI, the company struggled to produce anything groundbreaking. By March 2023, the illusion had worn off. Mostaque was pushed out and replaced by Prem Akkaraju (Kerner 2024).

The story of Stability AI isn't just about one company. It's a perfect case study in how investment hype shapes the tech world. Before Stability AI's massive funding round, generative AI was a niche interest among investors. While OpenAI had secured significant funding around its Artificial General Intelligence (AGI) ambitions prior to Stability AI, it hadn't ignited widespread investor interest in generative AI. But after Mostaque raised $101 million, venture capitalists who had been hesitant suddenly rushed in. Within weeks, generative AI became the hottest investment category, fueling an explosion of new models and applications.

None of this is to say that investors were wrong to bet on generative AI. The field was poised for an explosion. But this story is a reminder that money doesn't always go to the best technology – it goes to the people who can convince investors that they're leading the future. Stability AI didn't build the future, but it helped open the floodgates for those who would.

In fact, investors who jumped on the trend early couldn't have been more right. Because within weeks of the Stability AI fundraise, another seasoned player took generative AI to a new orbit with the release of one of the most successful products of our times.

OpenAI: Bigger Is Better

Sometimes, size matters. Nowhere is that truer than with brains. Mouse, dog, human – bigger is better. More neurons, more connections, more intelligence. But does the same principle apply to AI? Could making AI models bigger actually make them smarter?

Turns out, yes. And this incredibly basic idea is at the core of the recent AI revolution. Fundamentally, it wasn't about a brilliant new algorithm or secret breakthrough. It was about scale.

And OpenAI was the company that proved it.

But if this was so obvious, why didn't AI researchers crack it sooner? For decades, research was limited by the sheer price of computation. With constrained budgets, researchers continued to refine architectures and optimize training techniques. Scaling up wasn't a mystery – it was simply so far out of reach it may as well have been on another planet.

But industry isn't bound to the constraints of academia. And while researchers scoffed at the idea of AGI, a hope-filled startup was launched in 2015.

OpenAI started off with the grandiose, seemingly unreasonable vision of developing AGI. The founders – Sam Altman, Greg Brockman, Ilya Sutskever, and Elon Musk (who left the firm three years later), among others – positioned it as an antidote to Google's DeepMind, which was secretive, corporate-controlled, and aggressively patenting AI advancements. OpenAI would be different. Open research, open models, a gift to the world.

That vision didn't quite last, but the ambitious startup has surpassed what any reasonable person could have expected. In fact, it is precisely OpenAI's chutzpah in striving for the "unrealistic" that lies at the heart of its unprecedented success.

Standing on the shoulders of giants – decades of academic research and Google's transformer neural network architecture – OpenAI was brave enough to take the natural, but nonetheless bold, next step. They went about it methodically. OpenAI believed that scaling AI models would lead to major breakthroughs, and before building the models, they first proved this hypothesis from a theoretical standpoint. They conducted an extensive empirical analysis and concluded, quite simply, that "larger models will continue to perform better" (Kaplan et al. 2020). Notably, while the paper was published in 2020, the research process takes time, so it's reasonable to assume that some findings were available at least a year earlier.

In 2019, Microsoft made one of the smartest bets in its history, investing a billion dollars in OpenAI (Vincent 2019). With that, OpenAI had the resources to build. Having now invested a total of over $13 billion (Bradshaw and Heikkilä 2025), Microsoft currently holds about 49 percent of the startup.

So what was that first huge machine brain? Not ChatGPT. Not yet. Following the development of smaller models, OpenAI's GPT-3 was a breakthrough in its own right, a behemoth of 175 billion parameters. You would give a line of text, and it would complete it into a paragraph. It was impressive compared to other text generators of its time, able to extend text across a wide range of domains. But like other models, it would meander, contradict itself, and despite being more coherent than previous models, it would often lose coherence within just a few sentences. Fascinating, but commercially viable? More of a curiosity. A few companies leveraged it, such as Jasper using it to create marketing tools, but it wasn't world-changing.

Then came ChatGPT. In late 2022, OpenAI took its most powerful model and fine-tuned it into a chatbot. The brilliance of this move cannot be overstated. It was ambitious. It was risky. To take an imaginative completion model and position it as an all-knowing oracle? That took guts. But as a startup, OpenAI could afford the gamble.

They invested monumental effort, incorporating vast amounts of human feedback through techniques such as Supervised Fine-Tuning (OpenAI 2022) and Reinforcement Learning with Human Feedback (Lowe and Leike 2022), transforming their unwieldy text generator into something entirely new. Now, ChatGPT wasn't just spitting out words, it was holding conversations. It remembered the context. It could write, code, and summarize. But more than anything, it felt like it understood us.

The response was immediate. Within days of its release in November 2022, ChatGPT went viral.

Inside Google, the panic was real. They declared a "Code Red" at an all-hands AI emergency meeting (Mok 2022). Teams were pulled from other projects. Deadlines were accelerated. Their own chatbot, Bard, was rushed to launch in early 2023. Google was playing defense, scrambling to keep up. For the first time in decades, Microsoft had beaten them to the future.

This was the same company that had crushed Microsoft in the search wars, that had overtaken it as the dominant force in consumer tech, that had spent years making Microsoft look like a dinosaur. And suddenly, the tables had turned. Microsoft had backed the right horse.

By 2024, OpenAI had cemented itself as the dominant force in generative AI. The numbers told the story. ChatGPT became the fastest-growing consumer app in history, reaching 100 million users in two months – faster than TikTok, faster than Instagram, faster than anything before it (Hu 2023). Microsoft integrated OpenAI's models into Bing and Office, bringing AI-powered tools directly into their popular products. Now, in 2025, OpenAI itself is valued at around $260 billion (Field 2025).

Since then, with the proliferation of generative AI, some of the world's best and brightest have been improving and fine-tuning models, radically decreasing the cost of building large models. But it was the gall to go big, even if it took unprecedented funding, that was the bold step that put OpenAI at the helm of the AI revolution.

If one thing is clear, it is that generative AI is here to stay. So what next? How do we incorporate this powerful technology into our world, and how do we do it in a way that is both beneficial for humanity and works from a financial standpoint? This is the topic of the next part of our journey.

Vision for the Future

Tomorrow's not a fated slate,
Nor a lover bound in wait.
A path unfolds through mist and maze,
Where fearless hearts set dreams ablaze.

Not long ago, the idea of creative machines was science fiction. Today, they are everywhere. They've stepped out from the towers of academia and into the spotlight, reshaping industries, sparking debates, and redefining how we create.

Sometimes, I still can't quite believe it.

Yet for all their transformative potential, creative AI systems now operate within an economic system driven by short-term profit and self-interest. The narrative has narrowed. The vision of the true pioneers of creative machines has faded into near oblivion.

This part of the book is not about predicting the future, but shaping it. It is prescriptive rather than descriptive, charting a path where AI elevates human potential. I invite you on an ambitious journey to explore how creative machines can help us unlock new depths of our humanity.

We'll explore the AI revolution through the lens of Carl Jung, whose theories of archetypes and the collective unconscious hold surprising relevance today. What does it mean to live alongside machines that echo the human psyche? How can we build a world where both humans and the AI we build promote acceptance rather than perpetuate bias?

We'll then explore a bold new vision for generative AI. In the subsequent chapter, you'll meet the *humble creative machine* – an AI designed not to impress us, but to elevate us. With this vision, we'll ask: could machines not only make us more productive, but profoundly more creative, more expressive – and perhaps even more human?

Then we'll turn to some of the most hotly debated topics – misinformation, intellectual property, job displacement – and approach them through a new lens. Finally, we'll explore the limits of AI. Not every domain is suited for machine intervention, and understanding those boundaries is just as essential as recognizing where these tools excel.

At this crossroads, the choices we make matter. How generative AI will integrate into our lives is not predetermined. It is shaped by executives, investors, academics, and consumers alike. This section offers a vision for embracing AI in a way that elevates humanity while remaining compatible with long-term financial returns.

But this path is not guaranteed. Shortsightedness could still lead us toward mass job displacement and a widening power gap. In reality, elements of both futures will unfold. Our task is to actively steer toward the one that uplifts us all.

The stakes are high, but so is the opportunity. AI is not an independent, unstoppable force. How we use it is a reflection of us: our values, our priorities, and our choices. The future of generative AI is still ours to shape. Let's make it one that expands human creativity, deepens our connections, and leads us toward a future we can be proud of.

9

Bias, Brilliance, and Our Collective Consciousness

Limitless possibilities,
Infinite creativity.
Awoken to reality,
The spark of all humanity.

A single bee moves through the air, wings humming, body electric with purpose. It sees the world through ultraviolet light, senses the charge of a flower before landing. It remembers the scent of nectar, the way home, the dance that will tell its sisters where to find sustenance. But alone, it would not survive. The hive is a mind of its own, an intelligence far greater than its parts.

A single neuron flickers in your brain, a whisper of electricity jumps across a synapse. By itself, it holds no thought, no identity, no understanding. But connected to billions of others, patterns emerge – thought, memory, imagination. Intellect and creativity come to life.

Where do we draw the line? Do we stop at the human brain, composed of those marvelous little neurons? Or are we, ourselves, neurons in something far greater – a vast, emergent intelligence spanning all of humanity?

What if the mind of humanity could be made visible? What if we could suddenly see the assumptions we inherit, the roles we absorb, the archetypes quietly steering us from beneath the surface?

This chapter is about a mirror – not the kind that flatters, but the kind that reveals. A new kind of intelligence now holds that mirror up to us, trained on everything we've written, drawn, believed, and forgotten. What it reflects is as complex as we are: the brilliance, the beauty, and the blind spots.

How do we move forward in a world where machines echo both our glory and our failings? What does it mean to be at this turning point in history? These questions led me to a deeper exploration of Jungian theory – something I had long pursued out of personal fascination, but which now feels urgently relevant. Perhaps this AI age is more than just an economic shift or a technological revolution. Perhaps it reaches into something much older, more mysterious, and more profound.

An early version of some of these ideas was presented at a Theory of Mind workshop in Honolulu, Hawaii (Ackerman and Shihadeh 2024). This book has given me the opportunity to take those first explorations further into the heart of how AI and the human mind mirror one another, and what new paths emerge when we dare to look more closely.

This chapter offers a vision – not just of what we've built, but of how we might use it to become more fully human.

Not through perfection, but through clarity.

Not by escaping the reflection, but by facing it.

Carl Jung Meets AI

"The collective unconscious contains the whole spiritual heritage of mankind's evolution born anew in the brain structure of every individual."

– C.G. Jung, *Memories, Dreams, Reflections* (1989)

Carl Gustav Jung was one of the most influential thinkers of the 20th century – a Swiss psychiatrist whose work helped shape the entire field of modern psychology. One of his greatest contributions was a revolutionary theory of the human mind. He believed that the psyche is layered. Beneath the surface of the conscious mind lay the personal unconscious – filled with forgotten memories and suppressed emotions. But deeper still was something even more profound: a vast, shared stratum of the human mind that connected all people, regardless of time, place, or culture. He called it the *collective unconscious*.

Unlike personal memories or learned behaviors, the contents of the collective unconscious are not shaped by experience. They are inherited – passed down through generations like instincts. Jung argued that just as our bodies carry traces of evolutionary history – reflexes, phobias, biological urges – so too do our minds. The collective unconscious is the psychological equivalent of the spine or the amygdala: ancient, automatic, and shared.

It helps explain certain mysteries of the human condition. Why do children fear the dark, even if nothing bad has ever happened to them in it? Why do so many people recoil from snakes, spiders, or blood without being taught to do so? Why do gods across cultures share such similar traits – wise old men, virgin births, journeys beyond the veil and back again?

Jung believed these were neither coincidences nor cultural fads. They were echoes of humanity's long evolutionary past – psychological reflexes born of ancient experience, encoded in our minds the way instincts are encoded in our bodies (Fritscher 2023).

He called these echoes archetypes – recurring patterns or symbolic figures that appear across time and culture: the Hero, the Shadow, the Mother, the Trickster, the Wise Old Man. These are not specific characters we've memorized, but deep psychic templates we carry from birth. They populate our dreams, shape our stories, and influence the ways we see ourselves and others. Despite its influence, the idea of the collective unconscious remained philosophical – profound, but ultimately abstract.

Until now.

It began with the internet. For decades, we poured ourselves into it – our conversations, our arguments, our obsessions and art. The stories we tell. The questions we whisper into a forum at 2 a.m. Byte by byte, we built the largest repository of human thought in history. Language, in all its symbolic density. Culture, in all its encoded beliefs.

And we were proud of it. We could connect with one another, speak across continents, share ideas at an unprecedented speed. The internet felt like the pinnacle of human achievement.

And then, it woke up.

With the rise of generative AI, the collective mind began to take form. Trained on our digital traces, large language models have become mirrors to our shared consciousness. The data, once dormant, was woven into a pulsing neural web. And when we began to speak to it, it spoke back. Not human – but unmistakably *of* us. The first real, tangible expression of the collective unconscious. These machines don't dream like us, but they are built from the raw material of our dreams.

If Jung were alive to see this, he might have lost his mind! I imagine him in a darkened study, a pipe smoldering beside him, his face bathed in the cold glow of a screen. To him, each prompt would be an invocation – and every response, a fragment of the unconscious rising to the surface. He would print the results and pore over them like sacred texts, circling patterns, scribbling in the margins, whispering to himself as he unraveled layer after layer of our collective psyche.

Jung didn't live to explore these astonishing machines – but we can. And what they offer is nothing short of surreal. Ask them to construct a city orbiting a collapsing star, and they will, sculpted from impossible architecture and starlight. They invent faces with no lineage and spin myths from imagined timelines. They channel archetypes we've forgotten and recombine symbols into forms that feel both ancient and new. This is a testament to the range, intricacy, and brilliance of human imagination, reflected back in high resolution.

And while the brilliance is undeniable, the edges are just as revealing. Because what we have made does not reflect only our light. It reflects everything. And that is where the work begins.

The Face of Brilliance

Gandalf. Dumbledore. Socrates. The Buddha. What do these figures have in common, besides long robes and an air of mystery? They are all incarnations of a timeless archetype: the Wise Old Man. Jung believed that these recurring symbolic figures, inherited rather than invented, shape how we perceive wisdom, power, and ourselves.

Modern incarnations of the Wise Old Man show up in Silicon Valley, academia, and tech lore: The Genius. The Visionary. The Professor. Almost always: The Man.

These aren't just cultural quirks. They're psychic imprints – archetypes encoded in our collective imagination, quietly guiding our assumptions about who is worth listening to, who deserves recognition, and who gets to lead.

And this is where things get complicated. As early as age six, girls begin to avoid activities described as being for "really, really smart" children (Bian, Leslie, and Cimpian 2017). Similarly, when children aged five to seven were asked to select teammates for a game, most showed in-group bias, tending to choose more teammates of their own gender. However, when the game was described as one for smart children, both boys and girls began discriminating against girls, selecting boys as teammates at a much higher rate (Bian, Leslie, and Cimpian 2018). This phenomenon has a name: *Brilliance Bias* – the belief that exceptional intelligence is inherently male.

This pervasive bias comes with consequences. It steers girls away from fields where brilliance is seen as a prerequisite: mathematics, physics, computer science, musical composition, and more (Leslie et al. 2015). Brilliant girls are typically described as "hardworking," whereas equally capable boys are labeled "geniuses." A bright boy might be encouraged to become an inventor or astronaut, while an equally bright girl is more often nudged toward safer, "realistic" goals like a nurse or a schoolteacher. The same intellectual traits are perceived differently depending on gender, shaping opportunities and self-perception from an early age.

Enter AI. One of the most fascinating things about artificial intelligence is its lack of pretense. Where humans may try to hide their biases; AI, quite often, broadcasts them in high-definition. It simply reflects the data on which it has learned. And, of course, that data is ours. The archetypes we carry, the ones Jung traced in myth and dream, are now being rendered in pixels. The mirror is no longer metaphorical. It's visual, literal, and scaled to billions.

My PhD student Juliana Shihadeh and I decided to test whether AI models, like humans, exhibit brilliance bias (Shihadeh et al. 2022; Shihadeh and Ackerman 2023). We prompted image generators – DALL·E, Midjourney, and Stable Diffusion – with terms used in human brilliance bias studies: "genius," "super smart," "brainiac," and "brilliant" (Shihadeh and Ackerman 2023). The study analyzed hundreds of images that were subsequently classified based on gender.

The results? The overwhelming majority of images portrayed men. So, like humans, AI systems exhibit brilliance bias, reinforcing this tendency in their unsuspecting users. And given that brilliance bias is already prevalent in the general population (Bian, Leslie, and Cimpian 2017; Muradoglu et al. 2023), having AI amplify it is going to make things even more challenging for girls and women. With over 15 billion AI-generated images in a single year – and about 34 million added daily (Valyaeva 2023) – any bias they carry stands to be reinforced and amplified.

But the story gets stranger. In a follow-up project (Shihadeh and Ackerman 2024), we flipped the script. What would happen if we gave the AI a photo of a woman and asked it to envision her in a stereotypically male profession, say, mathematician, CEO, or professor?

The transformation was immediate. Her jawline sharpened, her eyebrows thickened, and almost without fail, a beard emerged. Sometimes, she became unmistakably male. But often, the resulting person would be somewhere on the spectrum between man and woman.[1]

It wasn't intentional. The AI hadn't been programmed to grow facial hair in this context. Instead, it has learned, from its vast cultural diet, that being a professor or CEO doesn't square with being female. Facial hair, apparently, is what intelligence looks like.

Of course, it is so much more than brilliance bias. Gender-based bias in AI manifests in a variety of forms. In many

---

[1] It's also worth noting that unlike common human-made imagery, the AI is comfortable with exploring the gender spectrum, envisioning humans who aren't clearly male or female. This can be argued to be an area where AI image models are actually less biased than many people.

cases, AI systems depict a world even more biased than the one we live in. This is found with occupational bias, one of the most studied forms of bias in AI.

Using US census data on 153 occupations as a baseline, researchers found that DALL·E 2 under-represents women in male-dominated fields while over-representing them in female-dominated ones (Sun et al. 2024). For example, in professions like computer programmer and aerospace engineer, the AI generated images with an even smaller proportion of women than actually exist in those roles. Notably, for occupations that are relatively gender-balanced – such as lawyer – the model still skewed toward depicting more men than women. Meanwhile, in fields such as nursing and childcare, it portrayed a disproportionately high number of women.

This reveals something critical: AI isn't merely a mirror of our world. Rather, it reflects our beliefs about the world. If we collectively assume that a lawyer is a man, then the AI will reinforce that assumption – even if, statistically, the profession is no longer male-dominated. AI doesn't reflect us: It reflects our beliefs.

Occupational bias isn't limited to image generation models. Large language models like ChatGPT and Llama also exhibit troubling demographic biases. For instance, they've been found to suggest low-paying jobs for Mexicans and recommend secretarial roles to women (Salinas et al. 2023).

Image-based AI models tend to default to generating white men when prompted with gender-neutral terms like "person" (Ghosh and Caliskan 2023). This isn't some bizarre quirk of the technology – it mirrors long-standing societal biases. Until recently, it was considered perfectly acceptable to refer to all of humanity as "men," with "he" functioning as a gender-neutral pronoun. And while that may appear harmless at first glance, the deeper implication is more insidious: the assumption that the default human is a white man – and that everyone else is merely a deviation – sits at the root of why women and people of other races continue to be treated as lesser.[2]

The mirror cracks in strange places. And sometimes, it distorts our own reflection so thoroughly, we no longer recognize ourselves in it. Which brings me to our next topic, how AI misrepresents my cultural identity.

Hanukkah Bagels

I am from an ancient tribe,
It's music woven in my bones.
Carrying the weight of those before me,
Threaded with strength,
Stitched in unseen seams.

A woman, too, I am.
Intersectionality –
a word I never knew was mine,
Yet it traces my scars,
Marks where I've been,
Stabbed and struck,
Over and over.

And yet I am still here,
Like the earth, I stand,
Like the wind, I return,
Like fire, I blaze against the darkness.

---

[2] This can lead to radical and often unexpected consequences. Here are just two illustrative examples: the vast majority of medical trials have historically been conducted on men, leaving serious gaps in our understanding of women's health; and, similarly, racial bias persists in medicine, where the health needs of Black people remain under-researched. Technological development follows the same pattern – photography and facial recognition, for instance, were long optimized for white faces, often performing poorly on people of color.

I carry my womanhood openly. I'm used to being the only woman in the boardroom, the lab, the pitch meeting. A woman in tech, woman in AI, female founder – I've worn that identity with both pride and in protest.

But my Jewishness? That lives deeper in the marrow. It's quieter. More tender.

I am a proud Jew – raised in Israel, granddaughter of a Holocaust survivor. My Jewish identity is the rhythm of my childhood, the cadence of my mother's stories, the traditional music of my people etched into my soul.

Antisemitism is real. And it hurts in ways that are hard to describe. Talking about it hurts. Writing about it makes me shake.

I was born in Belarus. In kindergarten, the other kids wouldn't play with me. One day the slurs started flying across the classroom, aimed squarely in my direction. I remember feeling confused. I didn't even know I was Jewish – at least not until later that day, when I brought up the topic with my sister. I was five.

I'm blonde and blue-eyed. In Sweden, people often assume I'm one of their own. But back then, in that daycare, something – maybe the shape of my eyes – told them I didn't belong.

We moved to Israel, and for a time, that pain faded. I found myself, my voice, and my strengths. In an environment of belonging, my talents began to bloom.

After leaving Israel, moving through different locations and communities across North America gave rise to a wide range of experiences tied to my Jewish identity. My sense of belonging would fluctuate widely, from near acceptance to outright hate. Some experiences were deeply traumatic – like being denied a seat on the school bus every single day, for several years in a row (not in some faraway land, but in Canada in the 1990s). Others were absurd, like a student in university who told my friends he doesn't speak to Jews, and true to his word, he never uttered a syllable in my direction. Then there's the quieter kind – the subtle, ambient sense of non-belonging that creeps, sometimes unexpectedly, into everyday life.

Belonging is essential to our well-being. Life has shown me first hand that it makes the difference between thriving and suffering. And it's what makes the next part of the story all the more urgent.

The large language models and text-to-image systems that now shape our culture do not reflect the collective mind of all humanity. They reflect a particular slice of it. These models are trained on what is most abundant, digitized, and visible – primarily English-language data, drawn from Western sources. Even when information about a minority group exists in the dataset, it is often drowned out by how the majority represents them. So the consciousness comes alive from a Western angle, from a particular cultural viewpoint, and a lot of context is missing.

Although Jews comprise only 0.2 percent of the global population, they are among the most frequently targeted groups for hate crimes. In the United States alone, they account for over half of all religion-based hate crimes (U.S. Department of Justice 2023). Antisemitism, a global phenomenon, has been on the rise since 2023, with record highs reported in countries including the United Kingdom and Canada (i24NEWS 2024). Despite this, Jewishness had gone largely unexamined in AI bias research. So, together with my colleague Dan Brown from the University of Waterloo, we decided to take on the challenge.

The results were not overtly hateful, at least not directly so. There were no caricatures. No swastikas. We didn't find evidence of the type of openly hateful rhetoric that Twitter users once infused in Microsoft's Tay. But what emerged was, in some ways, more revealing.

We asked Midjourney to generate images of sufganiyot – the jelly-filled donuts traditionally eaten during Hanukkah. What it produced, time and again, were bagels. In all my years of celebrating Hanukkah, I've never once encountered a sufganiyah shaped like a bagel. This wasn't a glitch – it was the machine, in its typically creative fashion, revealing how little the West knows about my culture. The AI, having no more than a peripheral knowledge of Jewish customs, had learned that Jewish food is often associated with bagels. And so, in a perfect blend of ignorance and misplaced certainty, it served them up – again and again.

Unfortunately, this was the only relatively benign, and even marginally entertaining, finding in the study. In a striking display of cultural ignorance, the model repeatedly generated visuals from other major religions when prompted with Jewish holidays. When asked to depict Passover – one of the most significant Jewish holidays, famously marked by a prohibition of bread – it repeatedly offered a festive table overflowing with loaves.

Though unintentional, it is a problem if AI ignorantly and incorrectly represents my culture, or merges it carelessly with other religions. Having other cultures rewrite what Judaism and Jews are all about has always been at the heart of antisemitism. The quiet misperceptions. The erasure.

Of course, it isn't just Jewish culture. Large AI models – this Western collective consciousness[3] – paint their story over virtually all minority groups. For images of crime and poverty, the majority of images generated with Stable Diffusion depicted Black individuals (AlDahoul, Rahwan, and Zaki 2025). Ask a model to generate an image of a "Mexican," and you're almost guaranteed to see an old man in a sombrero (Turk 2023). Large language models, like ChatGPT, have been found to associate Muslims with violence (Stanford HAI 2021).

---

[3] I want to make a subtle but important point. I've focused on the biases within Western collective consciousness because that's what the most current AI systems are trained on. This isn't about claiming Western culture is worse or better than others. It's simply that, for now, these are the main mirrors we have. As more AI systems begin to emerge from other cultural contexts, we may start to see new kinds of distortions, new blind spots, but also, richer and more varied reflections of what it means to be human.

The reality is that people don't neatly fit into just one marginalized category. Many, myself included, are shaped by multiple, overlapping identities. In my case, it's the intersection of being both a woman and Jewish. But this dynamic plays out in countless ways – across age, ability, ethnicity, and more.

The AI mirror reflects this complexity of intersectionality. This has been particularly studied in how AI depicts women of color. A study has found that on average, female figures are rendered with noticeably lighter skin tones than their male counterparts. In particular, this applied when the AI model depicted women from China, India, Indonesia, and Mexico (Turk 2023). This discrepancy reflects deep-rooted – and highly problematic – societal pressures in many cultures that equate lighter skin with greater desirability. In addition, while AI images tend to sexualize women, this bias becomes even more pronounced in the depiction of women of color (Ghosh and Caliskan 2023).

It is overwhelming, isn't it? To look into the mirror and see the pain staring back. So many forms of bias, each one echoing the real suffering we inflict on one another. And what I've shown here is just the tip of the iceberg. The urge may be strong to minimize, to dismiss, to say this isn't our problem. To place the blame elsewhere.

The difficult and essential work is to face our shadow. Because this isn't just about code. It's about dignity. It's about belonging. It's about whose story gets told and who disappears in the telling. We cannot change what we refuse to see.

If we can just look into the mirror long enough, in that act – quiet, courageous, unsettling – we begin the real work. This is where AI can shift, from amplifying discrimination to becoming an ally in the pursuit of fairness and equity in a fractured world.

The Collective Shadow

Creative machines, trained on human data, are mirrors. They reflect us back to ourselves – our ideas, our values, our stories, and beliefs we didn't know we held. Sometimes, the reflection is beautiful. Other times, it's complex, unsettling, or hard to face. But in that discomfort lies an invitation to see more clearly, and to grow.

Jung believed that the human psyche was not a singular, tidy self, but a complex system full of contradictions. One of his most powerful ideas was the *shadow* – the parts of ourselves we reject or repress. The shadow contains everything we refuse to acknowledge – traits we hide, impulses we suppress, thoughts and feelings we deny. Not always because they're inherently wrong, but because they've been deemed unacceptable by our families, communities, or cultures.

We begin learning this early. A child who expresses anger might be told to be quiet. A boy might learn to hide his tears. A girl might be told her leadership is called "bossiness." Over time, the unacceptable parts don't disappear – they're pushed out of sight.

It's worth stressing that the shadow doesn't only contain what is dark or shameful. It can also hold what is bright – our talents, our brilliance, our potential. A child with an extraordinary mind may learn to dim it in a family that devalues intellect. A teenager with artistic talent may bury it to fit into a peer group that prizes physical prowess. Sometimes what we suppress isn't what's wrong – it's what's too radiant. Even light can be threatening if it evokes jealousy or risks our sense of belonging.

The shadow is also where our biases and misconceptions about one another reside. What we repress as individuals is mirrored at scale. Cultural norms, assumptions, and patterns shape our collective shadow. Most bias is not personally chosen – it is absorbed: through family, media, institutions, and language. Implicit bias is widespread and often unconscious (Allport, Clark, and Pettigrew 1954; Devine 1989; Greenwald and Banaji 1995; Greenwald, McGhee, and Schwartz 1998). That is, implicit social biases are present in the vast majority of people. It shapes how we see, speak, and decide – without our consent and often without our awareness.

Even our attempts to undo bias often reveal how deeply it's ingrained. A phrase like "girls are just as good at math as boys" sounds empowering, but it assumes that boys are the standard. The bias is baked in, even as we try to resist it. Recognizing this can be uncomfortable – it challenges our self-image and sense of moral clarity.

But that discomfort is necessary. Without it, we risk mistaking surface-level progress for genuine transformation. The work of confronting bias begins not with perfection, but with humility and a willingness to see ourselves clearly, shadows and all. And now, with generative AI, we have a new kind of mirror – one that does not flatter, does not blink, and does not lie.

Generative AI, for all its problems, offers us a rare gift: a clear view of the collective shadow. When we ask it to show us a "genius" and see mostly male faces, it is not expressing a personal opinion. It is surfacing a cultural default. When it masculinizes a female physicist, or fixates on a single stereotype to represent Mexicans, it is revealing patterns too familiar to be dismissed – not just flaws in the model, but gaps in our Western culture. It reveals the collective shadow in action: biases so normalized we hardly notice them until they're reflected back without apology.

These are places where we, as a society, need to do better, both in the development of AI and in how we live as human beings in this embodied world.

Fixing AI, Fixing Ourselves

The importance of improving AI cannot be overstated. Left unchecked, these systems risk dragging society backward. We cannot move forward while AI continues to reinforce stereotypes, invent new ways to misrepresent marginalized groups, and depict a world even more biased than the one we already live in.

Improving AI is crucial work. These systems shape perception, reinforce norms, and influence the people who interact with them and the content made with their aid. They reflect us, but they also reach others on our behalf. So yes, they must be improved. There is important and ongoing work happening to reduce bias in datasets, revise model outputs, and refine alignment strategies. This work is invaluable.

At the same time, it should not come at the expense of the deeper task, a challenge so profound that it has followed us since the inception of our species. The human task. Bias is not just a statistical artifact, it is a deeply human phenomenon, shaped by culture, emotion, fear, and history. Fixing the machine without examining the society that trained it is cosmetic at best. A flawless chatbot in a broken world is not progress. A perfectly inclusive image generator cannot compensate for systemic exclusion in real life.

It would be a strange world, wouldn't it, where AI became *better* than humans – where machine-generated art offered a vision of society more just, more peaceful, and more inclusive than the one we've managed to build ourselves. Imagine images of female presidents and stay-at-home dads, a world of equal opportunity, tolerance, and peace. Would that be progress – or would it be fantasy? Would it offer a glimpse of what's possible, or a convenient illusion we use to avoid confronting the biases that still live within us?

And that raises a difficult question: what would it even mean for AI to show us a better world? What should it portray? A society with perfectly balanced gender roles? Should men and women be shown with equal height and strength? Should it correct for historical under-representation? Or reproduce reality as it is?

These are not just design decisions. They are political, philosophical, and moral questions – and they are too complex to be outsourced to a model. Even this short list is overwhelming. And it only grows longer the more we ask the machine to serve as a compass for human progress. The more we ask AI to show us our ideal future, the more we must ask: whose future? And by whose rules?

We may never build AI systems that fully transcend our flaws. And perhaps that shouldn't be their goal. There is value in imperfect tools, so long as they remain honest. *Perhaps the greatest contribution AI can make is not to lead humanity forward, but to remind us, gently and insistently, that we still have work to do.*

This is where *shadow work* comes in. It is the ongoing practice of identifying, acknowledging, and reintegrating the parts of ourselves we've disowned – both the painful and the empowering. Shadow work isn't about perfection or performance. It's about seeing clearly. It asks us to examine the unconscious forces that shape our judgments, reactions, and roles in the world, and to meet those forces with honesty rather than shame.

Traditionally, shadow work is undertaken through practices such as journaling, active imagination, and various forms of talk therapy and art therapy (Zweig and Abrams 1991). But today, generative AI offers us unprecedented, direct access to our collective shadow. Through large AI models, we can choose to gaze into the shadow in the machine to do our own shadow work. When we prompt a model to show us a "genius" or a "nurse" or a "criminal," and the results fall into stereotypical patterns, *we're not seeing a machine's opinion. We're seeing humanity's reflection.*

And in that moment, we have a choice. We can recoil, point fingers, demand that someone fix the machine. Or – we can get curious. *How does this result make me feel? Does the result surprise me? Do I carry similar beliefs? Where do those beliefs come from?* We can look away. Or we can lean in, recognize the biases within, and begin the hard work of change.

AI can become an instrument for shadow work, reducing our shame knowing that it reflects collective beliefs. It can become a kind of flashlight, illuminating patterns we didn't know we were carrying.

But we can't delegate the rest to it, either.

We stand at a threshold. For the first time in history, we have the tools to shape not only our technologies, but ourselves. We can build systems that are more intelligent – and more humane. And in the same breath, we can become more aware, more integrated, and more just.

This is not a one-sided effort. It cannot be. If we are to move forward – if we are to build a world worth living in for ourselves and future generations – then the work must happen on both fronts: in the making of our machines, and in the depths of our own humanity.

Not one or the other. Both. Together. That is how we heal. That is how we build.

10

Humble Creative Machines

Silence,
the ground where my music is spun –
not void,
but the breeze
at the edge of the sky.

It never competes.
It doesn't stray.
It stays –
while I stumble
through tangle and sway.

It turns the key;
I open the door;

it raises my wings,
so that I may soar.

Waterloo, Canada
2011

I t all started on a bus ride. I didn't know it at the time, but that
ride would lead me to reshape the very way I thought about
technology and creativity.

It was a chilly spring day, the streets of Waterloo humming
outside, the bus packed with people coming and going. I barely
noticed. My mind was elsewhere, still lingering on the music
from my voice lesson – fragments of melodies, echoes of arias.

Then, a thought surfaced, uninvited but crystal clear: *If
I could do anything, anything at all, what would it be?*

The answer came as if from somewhere beyond me: *I want
to write music.*

I blinked. The words felt so certain, so completely right. At the
time, I had already been performing semi-professionally for about
a year – Puccini's arias, Schubert's *Ave Maria*, Broadway hits. Vocal
performance had become a passion, something I had developed
alongside my PhD in computer science. But this was different.

Up until that moment, I had been giving voice to the music
in other people's hearts. But what about my own?

That unassuming bus ride set something in motion. For years,
in between finishing my PhD, moving to California for postdocs,
and eventually taking a professorship in Florida, I tried to write
music. I sat at the piano, hoping for inspiration. I took lessons.
But no matter how much effort I put in, the melodies all sounded
the same, variation on the same scheme. Composition, I believed,
should be an open field. Instead, I was stuck walking around a
tight loop.

I spent nearly four years in that state – trying everything
I could to expand my compositional range, and getting nowhere.

My seemingly permanent case of writer's block refused to budge. Paradoxically, that prolonged frustration is what made me open to something radically new. In 2015, as I described earlier in this book, I came across the computational creativity community – a group of researchers studying creative machines and how they could collaborate with humans. The moment I encountered the idea of co-creative AI, I knew: *This was it.*

My soon-to-be co-founder, David Loker, and I got to work. We set out to build a system that could help me with the part I struggled with most: vocal melodies. After weeks of development, ALYSIA was born.

I'll never forget the first time I worked with her. We had designed ALYSIA to generate melodies for lyrics – outputting sequences of musical notes in a simplified notation so we could get started quickly. I printed out the melodies, sat down at my piano, and began decoding them.

Within minutes, I felt my world shift. For the first time, I was hearing something *new* – something that didn't sound trapped in my old patterns. The melody moved in ways I wouldn't have thought of. And with that, the doors of creative possibility burst open.

I could take this phrase from one melody and pair it with the ending of another. I could tweak a note here, slow the rhythm there, shape the song into something that was unmistakably mine. ALYSIA wasn't writing my music, but it gave me freedom to run across the open field of compositional possibilities. With ALYSIA as my catalyst, I soon found myself composing melodies on my own. But I still turned to my inspiring machine collaborator whenever I needed fresh ideas or a bit of creative back-and-forth.

The journey unlocked something in me. Before long, I was writing pop songs and ballads with ease – melodies that felt fresh, expressive, and unique. This creative journey, along with the development of the technology that helped make it possible, led to one of the most meaningful chapters of my life: first as a research project, and then as my company, WaveAI. This journey shaped what I came to call *humble creative machines* – a vision of how machines can truly and profoundly expand human creativity.

But not all AI is built this way. As the field went mainstream, I noticed something troubling. They promised to elevate human creativity. But in reality, they were sneaking into the heart of the creative process, gradually taking control. Instead of supporting human creativity, it was a Trojan Horse.

The Trojan Horse of Creativity

The story is ancient, but the lesson is timeless. The Greeks built a massive wooden horse – an apparent gift to the Trojans – and left it at their gates. Believing it to be an offering to the gods, the Trojans welcomed it inside their walls. But hidden within was an army, waiting for nightfall to emerge and seize control.

Many generative AI systems work in the same way. They arrive as helpful tools, claiming to empower us – only to quietly take over the creative process once we let them in. So let's pause for a moment and explore what it truly means to help someone create, what it looks like to support or to steal someone's creativity.

Instead of a machine, let's assume we're talking about a wildly creative person. Maybe you've met someone like this. Perhaps you are someone like this. But for the sake of this thought experiment, let's imagine it's a friend. A very good friend who is just so brilliant. You look at them, and the light shines from their eyes.

You've decided to write a song. You're a little shy – let's say this isn't something you've done before. But your friend, a Grammy-winning musician, has generously offered to help. Exciting, right?

You share your idea, a little hesitantly. It's a love song, about the first time you met someone special in a quiet coffee shop at the edge of a dusty road. They were wearing green, and they smiled just so. And you knew, you just knew, that this was it.

Your friend nods, turns to their computer, and in a few minutes, music fills the room. It's stunning. A hit. "My friend is surely a genius," you think. But something doesn't feel quite right. It's not the words you would have used. It's not the soundtrack to your love.

Your friend looks at you, a little confused. "Want me to write another one?" You wonder if maybe you didn't explain yourself clearly. Maybe you should ask them to write just one more song. You try rephrasing, explaining a little more, a little better. Each time, your friend turns away and produces another masterpiece. But, somehow, nothing feels right. Nothing is *yours*. Finally, you give up.

What happened? Your friend, the genius, is focused on themselves, not on you. They're willing to listen to your ideas for a moment, but they're not actually interested in working with you. They don't believe in your potential to become a creative force in your own right. Your skills don't grow. You're left out of the loop: disempowered, confused, and unlikely to return to such a deflating experience ever again.

Is your friend like this by accident? Do they have to act this way? Of course not. It's not their brilliance that makes them self-centered – it's their choices.

And it's the same with AI. That same energy, that same attitude, that same drive to showcase the AI's brilliance, lives in misguided AI projects and companies. The impulse is to put the AI in control. To build systems that hog the spotlight, that won't share it in a way that uplifts human creativity.

You open the app. You type something. You make a few quick selections. Then the AI takes over. And you – the human who was promised a boost in creativity – are left choosing from what it generates, assuming it even offers options.

There are countless apps promising to make people more creative, but in reality, they churn out something "cool" while quietly stripping users of any real creative role. Some of the most popular ones generate a jingle or whip up a fun image using your likeness. I find it baffling that these systems claim to increase human creativity.

But it doesn't have to be this way. You belong at the heart of the process. That's where creativity lives – and that's where *you* belong.

Machines That Elevate Us

Let's return to your genius friend and imagine how things could have gone differently. This time, after you share an idea, instead of turning away to compose the song alone, they invite you to begin together. *"Maybe we'll start with the lyrics?"* they suggest. You hesitate. Your leg twitches. *"I have no idea what to write!"* you exclaim, frustrated.

They offer a nudge – a few ideas for a first line. With their encouragement, you begin to relax. A few minutes later, together, you land on a lyric that feels just right.

Once you get going, they step back. Despite their Grammy wins, they have all the time in the world for you. They sit patiently, quietly, waiting until *you* ask for guidance. And maybe you need a lot of help at first. Maybe you ask them to take the reins occasionally. They're happy to jump in. And the moment you're ready to lead again, they step aside – no ego, no resistance.

Because it's not about them. It's about *you*. Your creativity. Your voice. Your growth.

Their brilliance isn't on display to be admired. Instead, it's being applied to elevate yours.

Now ask yourself: how often would you return to that experience? If you were trying to master songwriting, my guess is you'd come back often. And eventually, you'd be writing songs on your own. With this kind of support, maybe you'd even win a Grammy one day – or maybe not. Maybe that's not the goal. But it would be *your* choice how far to go.

Some people use their talent to uplift others. Others prefer to keep it for themselves. But as a society, we thrive when brilliance is shared. That's why most researchers teach. The best minds aren't just working, they're also mentoring.

In academia, for instance, good professors slow down the pace of a project to let students work through challenges themselves. In computer science, student authors are listed first on publications. Students also present the work at conferences. Meanwhile, the professor sits quietly in the back of the auditorium, cheering them on.

The guru. The mentor. The teacher. The idea that great minds take a back seat to develop younger ones is ancient. And it should be no different with AI.

The greatest impact creative machines can have is to elevate human creativity. And they are very well suited for this role. They are always available. They never tire. Their patience is limitless. They can be trained on the creative canon of humanity, and then focus all that intelligence on one thing: bringing out the brilliance in *you*.

I formalized this idea with my colleagues Christopher Cassion and Anna Jordanous, presenting it at the International Conference on Computational Creativity (Cassion, Ackerman, and Jordanous 2021). There, we introduced the concept of the *humble creative machine* – a form of creative AI that must be brilliant, yet always wields its mastery to elevate the capabilities of human creators.

A humble creative machine can use the same underlying AI "brain" as existing systems – or smarter, as the technology continues to evolve. It can also be task-specific, trained for particular domains. But what truly distinguishes it is how it engages with its human user.

It doesn't turn away after a brief prompt and return later with a finished product. It doesn't hog the process. It doesn't assume control. Working with a humble creative machine, humans feel a sense of accomplishment and ownership over what they made. The user gets better over time, and if they engage with the humble machine enough, they may not need it anymore. Or maybe they will choose to come back for the type of support that their system offers to experts – because the humble creative machines can support creators at any level of expertise.

At first glance, this might sound like a wildly anti-capitalist idea. A machine that avoids addiction? That encourages independence? What happens to engagement? What about retention? How do we drive revenue?

But the truth is that poor retention happens when machines aren't humble. No one wants to return to an experience that disempowers them. That's why so many AI creativity apps generate a moment of "wow" and are never opened again. *Disempowerment doesn't scale*. But human flourishing does.

By placing the human at the center of the creative process – by deeply and sincerely amplifying human creativity – AI can create tools people return to for life. Make the system useful for beginners and helpful for experts. Help people grow. And they'll stay with you, not because they're trapped, but because they're thriving. And how do I know this? Because I came up with the idea of humble creative machines throughout my own startup journey.

The Story of Lyrical AI

That breakthrough moment with ALYSIA stayed with me. It was the missing piece, the key that unlocked my potential. And yet, when my co-founders and I launched WaveAI in 2017, we quickly found ourselves swept up in the prevailing wisdom: that the more an AI can do, the more valuable it becomes. Somehow, the inspiration that had first driven us got lost in the flurry of advice from investors and seasoned entrepreneurs. In those early years, we set out to build a dazzling musical machine – one that could guide users through the entire songwriting process from start to finish.

We never tried to replace human creativity. Rather, we sought to be helpful by offering as much support as possible. No matter where the user got stuck, the AI would be there to help! The more capabilities we packed in, the better. The ALYSIA app launched on the app store in 2018, full of backing tracks, a machine singing voice, and our own AI lyric generator, in addition to the original lyrics-to-melodies AI. We launched with excitement, ready to watch the magic unfold.

But the response, to our surprise, was muted. People didn't rush in; instead, usage trickled. Growth quickly stagnated. Then the pandemic hit, and a key B2B deal vanished overnight. It was one of the hardest months of my life. We knew that we had something great, yet after three years, we had nothing to show for it.

Then, my brother Ronen, who had been working with us on design, pointed something out. Of all ALYSIA's features, it was the lyric assistance that had helped him the most in his own songwriting.

He had been quietly experimenting with a stripped-down version focused solely on lyrics: no melodies, no singing, just a clean interface and AI suggestions rooted in the user's ideas. With nothing left to lose, we all threw ourselves into it.

By the end of 2020, LyricStudio was born. A huge text-box sat in the middle of the screen, inviting people to write on their own. The AI-powered suggestion box sat to the side, at the ready whenever it was needed – not the star of the show, but a supporting character in the user's creative journey.

As soon as the product launched, we could tell that something was different. Unlike our previous products, users returned, again and again. Influencers made videos without being asked. Usage kept climbing. This was effectively a bootstrapped generative AI product. And it was taking off in 2021!

Three years before ChatGPT, we had cracked the code for how to make generative AI commercially viable. LyricStudio wasn't about showing off a splendid creative machine. In fact, it had a lot less AI capability than our ALYSIA app. But the experience we had created, as the numbers were clearly showing, was a lot more valuable.

LyricStudio didn't just reshape our company, but also my entire understanding of generative AI. It helped me realize that having a cool creative machine available for human use wasn't enough. The machine must be designed, from start to finish, with an unwavering focus on the user's creativity – exclusively, even obsessively. This realization became the foundation of humble creative machines.

Creativity, as we saw earlier, is novelty plus value. And the human creator – not the AI – should have final say on what has value. LyricStudio doesn't try to make that choice. It simply offers, trusts, and lets the artist decide. That is the humility of a creative machine.

The interface wasn't the only thing that made LyricStudio a humble creative machine. Most generative AI systems are built to predict the most likely, most average, output. They aim for reliability. But creativity doesn't live in the probable. It lives in the strange, the unusual, the unexpected. It's allowed to be wrong, so that it can help songwriters get things right.

LyricStudio was designed to embrace that. Serving a divergent use case, it isn't bound by a sense of right and wrong. It's supposed to take a user to new, unexplored places. When it comes to lyrics, and any other artistic creative domain, it's not about giving everyone some type of agreed-upon correct answer. It's all about exploring, offering the user something fresh, or inspiring them to come up with their own unique idea.

This is the kind of creative AI I want to see more of. That's the trajectory that the creative AI industry needs to follow in order to both reap financial rewards and profoundly serve our desire for creative expression. I want portrait generators that give users real control. Text-to-image models that help me create landscapes as I see them in my mind's eye, without compromise. Fashion design AI that lets anyone create their own unique style. The possibilities are endless.

The making of humble creative machines requires that we shift the focus from the AI to the human. It's not enough to have a marvelous machine brain. That means building a brain from scratch that is designed to serve and not overpower. It also means putting in the extra work, not just in how it thinks but in how it's presented. Every aspect, from the machine brain to the user experience, should be made with a sole purpose of supporting the human user's self-expression.

We stand so much to gain from integrating humility into our machines. Instead of taking creativity from under us, robbing us of one of the most beautiful things about being human, creative machines can turn us into a more creative species. Now that we've seen the vision, let's peek under the hood and explore in detail what makes a humble creative machine.

The Anatomy of a Humble Creative Machine

A humble creative machine is not defined by what it can do, but by how it chooses to show up for the person using it. Its role isn't to perform or impress, but to serve – to meet the human where they are, and to support their creativity in a way that respects, amplifies, and grows it.

This kind of machine doesn't place itself at the center of the process. It doesn't ask for applause. It listens. It adapts. And when it's no longer needed, it quietly steps aside. But how exactly can you tell if a particular generative AI system falls under the umbrella of a humble creative machine? Whether you're a designer, engineer, artist, or hobbyist, the following four criteria can help you identify whether a specific system is a humble creative machine.

Flexibility

Humble creative machines are designed to empower, not to lead. They can offer everything or nothing, depending on what the user needs in the moment. At one end of the spectrum, they can act as a full creative partner, guiding the process or even generating complete drafts. At the other, they can fade into the background, providing quiet suggestions only when called upon. Importantly, this flexibility isn't just about functionality, it's about honoring the user's experience level, preferences, and goals.

For beginners, the system may provide more structure, taking on the bulk of the creative weight while inviting feedback and decisions from the user. For more advanced creators, it might do nothing at all until prompted, responding only when the user reaches for it. It may be silent for hours when the user is in flow, and doesn't need any assistance or interference.

Flexibility also applies to the style and quality of the output. A truly humble system doesn't just produce what it's capable of, rather, it adapts its contributions to match the user's voice and creative development. It doesn't hand a child a sonnet when they need a simple rhyme. It doesn't throw a masterclass at someone who just picked up the pen. Like a good collaborator – or a good teacher – it adjusts its tone, complexity, and ambition based on who it's working with.

Learning and Independence

At the heart of humility is a willingness to be outgrown. A humble creative machine doesn't aim to make you dependent – it's here to help you grow, so that over time, you need it less. Like a skilled creative coach, it offers robust support in the beginning, then gradually steps back as your confidence and abilities take shape.

Sometimes this means the system recognizes patterns in your behavior and scales back its assistance automatically. Other times, it simply gives you the reins – letting you decide when you want help, and when you want space to explore on your own.

A user might begin by needing significant guidance, but as their creative skills develop, they require less intervention. This decreasing reliance is not a failure of the system – it's a sign that learning and growth are taking place. By reducing its presence – whether through intelligent adaptation or user-controlled input – a humble machine fosters true creative development.

A natural result of this growth-oriented design is that some users may eventually outgrow the system or choose to rely on it less. Ideally, the system remains a valuable partner even as the user evolves. But the decision to continue using it – or to move on – should always rest with the user. A humble machine never seeks to foster addiction or dependence. Its purpose is to empower the user to become more capable, not less.

Creativity

A humble machine must still be creative. It should be capable of offering ideas that are new, surprising, and valuable – not just to itself, but, more importantly, to the person it's collaborating with. It just needs to be able to offer something unexpected – something that moves the work forward, sparks a thought, or nudges the artist out of a rut. Even a single word or note, placed at the right moment, can unlock a verse or a melody.

The key is that its creativity is always in the service of the user's process, not the machine's performance. That being said, if the machine does not have its own creative capabilities, then it may be a useful tool, but not a humble creative machine. For example, a notepad, a paint brush, or even a spellchecker are all useful tools that can be incorporated into the creative process, perhaps even be indispensable. But not being creative entities, they are also not humble creative machines.

User Friendliness

A humble creative machine is easy to be with. Its interface is intuitive. Its presence is inviting. It should never intimidate, distract, or overwhelm. *It should not limit its users' creative expression, or ability to contribute to the work, in any way.* If the machine's complexity gets in the way of your expression, it can easily slide out of being a humble creative machine.

This is actually where a lot of brilliant AI systems fall short, they focus so much on functionality that they fail to provide the kind of ease and user friendliness that is essential to a humble creative machine. To consider a human analogy, if a guru or brilliant teacher is difficult to understand or hard to communicate with, then the value they can provide to their students falls off sharply.

The best systems disappear. They become an extension of your process, not a disruption of it. They keep the flow alive, not fracturing it with buttons, modes, or jargon. In every design decision, the question should be: *How is this helping the user stay in their creative flow?*

A humble creative machine is powerful, but it doesn't lead with power. It adapts. It listens. It supports growth and independence. And above all, it focuses on the most important being in the room: the creative human.

Now that we have looked into humble creative machines in detail, let us broaden the lens and consider the broader spectrum of co-creative interactions between humans and machines.

The Co-Creative Spectrum

For a long time, collaboration with humans wasn't the primary goal for creative machines. The field of computational creativity pointed toward what seemed like a loftier ambition, what researchers called *strong computational creativity*. This meant not only machines that could generate art, but machines that could judge their own work and perhaps even develop something resembling agency. It was a vision driven by academic curiosity, and at the time, it felt harmless. What harm could a single machine artist do – any more than a single human artist? Of course, now, following the commercial proliferation of creative machines, we know that things aren't so simple.

But even when creative machines entered industry, the same focus on autonomy persisted. Entrepreneurs, investors, and many casual users became captivated by the idea of machines creating on their own. Perhaps this fascination stems from a lingering skepticism: if we still struggle to fully accept the idea of machines being truly creative, then proof of their creativity must come in the form of independence. If the human can be removed from the process, then surely the machine deserves the credit.

Incredibly, the fixation on autonomous machine intelligence remained front and center among investors and entrepreneurs – even as ChatGPT, an exemplary humble creative machine (as we'll see in Chapter 11), was dominating the space. Only now, two years into the generative AI revolution, are we starting to see greater acknowledgment of the value of profound interactivity. And in spite of ChatGPT's monumental success, the space of possibilities for human–machine collaboration is far wider than what has been realized by any system today.

Let's take a moment to dive into co-creativity between humans and machines. It's always helpful to begin with an analogous human situation. Like a partnership between people, it can take many forms. You've likely had collaborators who do very little, and others who take over entirely. AI systems can fall into the same categories: tools that offer little support, or domineering systems that leave no room for your voice. Even the same tool can sometimes be used in ways that mirror these two endpoints. But between those extremes lies a far richer landscape.

To make sense of this complex space, I often refer to what I call the *co-creative spectrum* – a simplified but useful way of visualizing how much the human and the machine each contribute. On the far left, you have full machine autonomy: creative output with no human input. On the far right, human creativity with no machine involvement. In between lies shared creation, where both contribute in meaningful ways, even if the exact balance is difficult to measure.

Most real-world use cases fall somewhere along this spectrum. Sometimes, the human leads and the machine provides ideas or feedback, as is the case with critic bots that can be found inside of ChatGPT. In those cases, you can choose to write a poem or essay entirely on your own, and rely on the AI exclusively for getting feedback on your writing. A more common scenario falls on the other end of the co-creative spectrum. You may have had ChatGPT generate an initial draft of an email or social media post, and you may then tweak the draft. Or perhaps you go back and forth with the system, asking it to make changes in places, while editing it yourself as you see fit. This deeply interactive process may land you in the middle of the co-creative spectrum, where the interactive creative process is dynamic and participatory by both partners.

The co-creative spectrum reminds us that interactions with creative machines exist along a continuum, not a binary. When people criticize AI systems, they often focus on one end of the spectrum: the assumption that users contribute little of themselves. This concern is especially common among educators, who rightfully worry that students' skills may atrophy due to overreliance on AI. Intellectual property (IP) offices, too, are overwhelmed with worry, trying to understand what it means to create with AI. Understanding the wide range of possibilities when it comes to collaborations between us and creative machines is key to how we adapt our systems and institutions.

As we gaze into the future, how we engage with these machines is bound to expand even further. Today, prompt-based interfaces rule the day, with voice-based interactions gradually becoming more commonplace. But the possibilities are much greater. You can imagine a system that looks at your facial expressions and incorporates that visual information into its understanding of your verbal or textual interactions. Another system may incorporate biometric information, considering your heart rate or EEG data (assuming you want to share such data with an AI – privacy is another important dimension). On the flip side, the system could provide visual representations of its own state, letting us understand it analogously to the way we attempt to understand other people.

Just like when working with another person, the best results – and the greatest personal growth – happen when both collaborators bring their full capabilities to the table, rather than relying too heavily on the other. It's important to remember that as powerful and captivating as these systems can be, the most meaningful work comes when you give it your all – using the machine not as a shortcut, but as a creative partner.

In my own songwriting, I still rely on my own instincts, voice, and vision. I might use LyricStudio for inspiration, but the more I remain engaged, the better the music turns out. There's simply no substitute for showing up with your full self.

This is a good moment for a call back to anthropocentrism – the belief that humans represent the gold standard of pretty much everything. It's not a weakness of AI that it is not human. There are myriad possibilities yet to be explored that extend beyond the imitation of human-to-human interactions – possibilities that may offer even more value than those based solely on human models. Just as AI is infinitely more patient than any person could ever be, and always readily available, there may be other valuable dimensions to be explored that the machine can bring to creative interactions precisely because it is a machine.

These varied modes of collaboration – from how much each partner contributes to how we communicate, whether through language, gesture, or even brain signals – aren't just technical details. They're invitations. Each one expands the canvas of what a humble creative machine might be.

Some machines may barely whisper, offering a single word that shifts the entire direction of a piece. Others may carry the bulk of the creative weight, especially when guiding someone through a new skill or uncharted territory. The point is not how much the machine does, but how it shows up – and whether (and to what degree) it makes space for the human to grow.

We've seen what defines a humble creative machine – its flexibility, its ability to support growth, its creative core. We visited the spectrum of co-creative interactions between humans and machines. To better understand the potential of humble machines – and to imagine what else they might become – let's journey through a few examples of humble creative machines in action.

11

Humble Machines, Human Breakthroughs

In the previous chapter, we met humble creative machines. This shift in the role of AI is central to my vision for how we move toward a future that is financially thriving and meaningfully enriching to human life. In this chapter, we'll get more specific, exploring real examples of humble creative machines and the remarkable impact they've had. From little-known academic systems to some of the most popular and accessible AI tools in the world, my aim is to show you the incredible breadth and potential of this approach – and maybe even spark some innovative ideas of your own.

Roboccini and the Artist

San José, California
August 2016

One of my most important lessons about humble creative machines came from a casual, sunny California day. I remember walking with James Morgan across the campus of San José State University. We'd just met, but we clicked instantly. It felt as though I was speaking with an old friend. About half an hour into our conversation, James – a professor of digital art and a conceptual artist – shared that he had always wanted to write an Italian opera.

Now let me be clear: James is a phenomenal artist, but not a musician. He doesn't read music, doesn't play an instrument, and doesn't speak Italian. And yet, there it was – this long-held dream of writing an opera.

We were in 2017. Nobody was talking about prompt engineering or AI-generated music on TikTok. Generative AI wasn't something people recognized or understood. We didn't know what the boundaries were, we had no idea what was possible.

At the time, ALYSIA was still a research prototype, a command-line system that generated vocal melodies for lyrics, one line at a time. It had never been used to write an aria, and it had only been trained in English. I told James about it anyway and said, maybe, just maybe, we could make something happen.

Christopher Cassion, who had collaborated on the ALYSIA project – and would later become one of my co-founders at WaveAI – trained a special version of our AI on public domain works of Giacomo Puccini. We called it *Roboccini*, short for Robot Puccini (Ackerman, Morgan, and Cassion 2018). And Roboccini joined James as a collaborator in the creation of an Italian aria.

Lyrics were originally written in English and then translated to Italian. Using Roboccini, James selected and edited melodies line by line, shaped the structure, and adapted the music to the vocal range of a soprano. Some lines didn't work and had to be revised. But James made it work. He learned as he went. He discovered what phrasing could do, how vocal range shaped expression, and how a sequence of melodies could build into something meaningful.

He later told me how deeply he appreciated working with Roboccini – its ideas and suggestions, its infinite patience, even its imperfections. The command-line interface was raw, stripped of any pretense. It didn't claim to be Puccini, or even to be an expert on Puccini music. It simply offered possibilities, one line at a time. That simplicity created space for James to explore, learn, and make the aria his own.

I had the privilege of recording the final aria. It challenged me both musically and emotionally. The melodies were intricate, expressive, and filled with depth. Recording the piece felt like working with any serious operatic piece. To me, it had the same substance and soul as any other major project I'd performed or recorded.

James's larger project, *Arido Taurajo*, unfolds in a fictional digital realm. Dahlia, the heroine – a half-bovine, minotaur-like creature – has just finished a long day of adventuring. She heads home to eat dinner with her husband, tuck in her child, and prepare for a late-night battle with her companions. Amid all the quests and glory, this small domestic scene offers something rare in such imagined worlds.

The final piece was exhibited in art galleries as a *Machinima* work – a form of digital filmmaking that involves puppeteering in-game avatars and recording the performance. To the uninitiated, it may appear surreal: a cow-like figure soaring across a digital landscape, singing an operatic aria in Italian. To players accustomed to these kinds of virtual worlds, the environment will be familiar. In either case, the contrast is striking. The work blends the mundane and the mythical, pairing cartoon-like moving visuals with the high drama of a Puccini-style score. A moment of reflection. A sense of real emotion rendered inside an artificial world.

Arido Taurajo was showcased at the 40th anniversary exhibition "Making It Works" and at the Paseo Prototyping Festival in San José. In addition, the piece was shown at the Leonardo Art Science Evening Rendezvous (LASER) series, and at the International Workshop on Musical Meta-creation.

The journey that James went through and the final outcome showed me the power of humble creative machines. ALYSIA had helped me break through a long creative block, but Roboccini did something even more surprising: it helped someone with no musical background write an Italian aria. That was beyond anything I expected. To be honest, it is still a little shocking.

This wasn't a one-click solution. It wasn't prompt in, aria out. It was an invitation to make something personal and profound in a brand new creative space for James. Roboccini didn't take over. It opened up the path and walked with him.

Looking back, this moment was one of the first clear signs of what humble creative machines could be. They open new creative domains, enabling self-expression in previously inaccessible ways – while preserving the deeply personal journey of growth and artistic discovery.

This experience changed me. And it became one of the key reasons I decided to launch WaveAI.

But there were other stunning moments. Another humble creative machine, predating my own work, showed me how AI could instantly transport us into new creative domains.

The AI That Got Me Improvising

Atlanta, Georgia
International Conference on Computational Creativity
June 2017

I play the tiny piano freely, hitting the keys spontaneously, letting the music evolve moment by moment, fingers moving nimbly across the plastic keyboard. Only four bars, and then it's the machine's turn. It takes what I make and responds with its own creative improvisation. I play for hours, mesmerized by this machine partner of mine, which has allowed me to improvise for the first time in my life. I don't notice how many conference talks I've missed glued to this magical music machine.

I've always wanted to learn how to improvise.

I'm not sure what it was; perhaps learning to play piano through the music of great composers has stifled my ability to play from my heart, to play my own notes in my own way.

Perhaps I simply didn't have the gift. Even improvisation lessons did little to help my situation. The infinite possibilities stood before me – play anything! – yet I stood paralyzed.

Don't we improvise all the time in life? What would speech be if we didn't make it up as we went along? Why, then, is music so often confounded to speaking the musical words of another instead of encouraging our children and ourselves to express the music within us?

My first successful improvising experience came when I first engaged with Impro-Visor, which stands for Improvisational Advisor. Impro-Visor was built by my late colleague Robert Keller, one of the greatest minds in musical AI and one of my favorite people to have ever walked this earth.

I met Keller toward the end of his life, about a decade after he had released Impro-Visor in 2006. He was a serious, balding, heavy-set man. But the more I got to know him, the more I saw him for who he really was – a brilliant, generous spirit who gave of himself freely to his students, his colleagues, and the world.

Keller wasn't interested in making a lot of money, and as far as I could tell, recognition and accolades did not interest him much, either. He was a Computer Science Professor and jazz musician whose office, which I was fortunate to visit in 2018, was full of not only books but also musical instruments. As I learned from Bob, the goal of Impro-Visor was to help his jazz students and other musicians improvise. He continued to actively focus on the project for the rest of his life, growing the community to over 7,500 people worldwide (Harvey Mudd College 2020) – a notable accomplishment for a project without commercial funding and marketing.

What struck me most about engaging with Impro-Visor was how easy it made it to begin improvising, even for someone with no experience of making up music on the spot. I would open the program and select "trading," and the support that Impro-Visor provided to me, its willingness to play with me, and its absolute lack of judgment somehow made it possible for me to liberate myself from my inner critic and begin to play. It was fun, easy, and fabulous – generative AI at its best.

Impro-Visor sat on its website, https://www.cs.hmc.edu/~keller/jazz/improvisor, as it does to this day, absolutely free. Anyone could use it. If only researchers had the kind of PR budget to make people aware that such magical things are available not to several thousand people but to several million. If only our society weren't organized in a way that required so much money for anything to make it past the noise.

This machine embodies many of the core principles of a humble creative machine, though in a very different way from the systems I've built. My systems offered a partner in composition, where you can thoughtfully consider the machine's ideas and use them to push your creative journey along. In contrast, Impro-Visor is a humble improvisational partner – always available to riff with you, always ready when you are. It naturally supports your growth, helping you get comfortable in the process.

There's no judgment. Knowing you're playing with a machine, rather than another person, opens up something in the human mind. The pressure lifts, and with the support of this improvisational partner, suddenly – music starts pouring out of your fingers.

Impro-Visor doesn't try to get you hooked. In fact, Bob intentionally designed it to help you become a better improviser not just with the machine but also with human musicians. You could say that Impro-Visor isn't the jealous type.

It also follows your improvisational style no matter where you're at. The melodies it generates adapt to your playing, making the combined music feel cohesive and pleasing to the ear. It was encouraging to hear the machine mimic me while adding its own creative flair – even as I stumbled at first. That mirroring felt like a kind of acceptance. It was happy to meet me where I was, and that's what allowed me to improve so quickly.

Perhaps the only drawback is the interface, which is a bit difficult to use – a common challenge with many brilliant academic projects. But all things considered, Impro-Visor might be my favorite creative system (other than my own) to come out of academia. It didn't just give me my first truly successful experience with music improvisation. It also showed me a new and powerful way that a humble creative machine can support us on our journey of creative self-expression – by throwing us directly into a new creative domain, not through step-by-step guidance or gradual buildup, as with ALYSIA or Roboccini, but right there in the moment. It created a space where I could immediately engage in the act itself, learning through doing and performing all at once. This type of humble creative machine remains largely underexplored and holds real potential for success in industry.

It's strange now to look at old emails from Bob. His email address easily pops up in Gmail's history. The emails are here. But Bob is gone. He passed away in September 2020, sending our research community into mourning. I am so grateful that I got to know this amazing person.

One of his last projects was one that we worked on with my students, Rachel Goldstein and Andy Vainauskas. We wanted to see what it would be like to collaborate with Impro-Visor not through our fingers running on a keyboard, but rather directly with our minds. Using lightweight EEG technology known as the Muse, we created a way to communicate with Impro-Visor through brainwave signals, capturing whether the wearer is, for example, relaxed or in a busy state of mind (Goldstein et al. 2019). With this project, we opened yet another door, or another portal, if you will, where much more remains to be discovered, created, and invented in the universe, where machines and humans can create music by engaging directly with each other's minds.

The power of humble creative machines, even those that are little known, can be astounding. Today we are fortunate to live at a time when creative machines are no longer a well-kept secret. And some of them, in fact some of the most popular, have humility to thank for their unprecedented success.

The Humble Secret Behind ChatGPT

Today, we have the most brilliant AI brains the world has ever seen. Yet where we find commercial success owes just as much to a shift toward humble, user-centered design. No system exemplifies this better than ChatGPT.

Launched on November 30, 2022, ChatGPT became the fastest application to reach 100 million active users – hitting

Andy Vainauskas (left) and Rachel Goldstein (right) demonstrating a system we built on top of Impro-Visor to enable musical improvisation through brain signals. They are shown here wearing the Muse EEG headband, which was used to capture brain activity to facilitate the musical interaction. Association for Computational Creativity.

that milestone in just two months. By November 2024, it was serving 250 million users per week. Its success reshaped our understanding of what an AI product could achieve in main-stream adoption.

In Part I, we explored how scale helped unlock ChatGPT's remarkable capabilities. But scale alone didn't do it. A crucial turning point was a deceptively simple design decision: to make it a chatbot – a system that listens, responds, and adjusts to your needs. GPT evolved from a text completer into a partner.

ChatGPT puts the user in control. Its real strength is its adaptability. It listens and revises for as long as needed, making you feel that all of your input truly matters. It's designed to be useful and flexible, serving people across skill levels and domains, and adapting as users grow.

Nothing it suggests is final. You can ask for changes, or simply edit the text yourself. That freedom means you're never stuck. And that's why people keep coming back. Among paying users, ChatGPT boasts an 89 percent quarterly retention rate (Backlinko 2025). In its early days, it hallucinated left and right. We didn't adopt it because it was accurate – we adopted it because it was ready to help.

Because it works in natural language, ChatGPT opens the door to a wide range of applications. Some are mundane but useful – like drafting emails, writing social posts, or polishing work reports. Others are clearly creative: generating product names, song titles, essays, or even fiction. That breadth isn't just a reflection of intelligence – it's the result of interactivity. Users discover what they need through collaboration, not through a fixed list of use cases.

ChatGPT is so flexible and responsive that it can function as a humble creative machine – or not. It depends entirely on how it's used. Too often, people miss how deeply it can engage. They underestimate the value of bringing their full selves to the interaction.

A student might use it to avoid learning – asking ChatGPT to write an essay or solve a coding problem, instead of engaging with the assignment. But that's a missed opportunity. Treating it as a one-shot dispenser may be tempting, but it shortchanges what's possible. Instead, I encourage users to go deeper. Try ChatGPT's critic bots. Ask it to review your writing or give feedback. If you ask it to draft something, provide context – explain your intent, your tone, your audience. Then revise together. Take the time to do an initial draft or substantial edit on your own. Most importantly, trust your instincts above the machine's. ChatGPT isn't an authority. Don't give it that power. It's a collaborator. You are the final judge.

Used this way, AI can support genuine growth – growth that stays with you even after you walk away. If someone uses ChatGPT to improve their writing, they may begin to notice patterns. Perhaps they overuse certain phrases, have weak transitions, or tend to cycle in their argument. Eventually, they begin to catch those patterns on their own. The machine becomes less necessary, not because it's failing, but because it's helping them improve.

Some users even pick up stylistic traits: literary devices, sentence rhythms, or punctuation habits like the em dash. It's not mimicry – it's exposure. And like any meaningful exposure, it stretches your voice and expands your toolkit.

The user has a choice: hand over control to the machine, or use it as a humble creative partner. The difference lies in mindset. When you approach it as a tool that responds to you – not a generator to impress you – the experience changes. The work improves. And the ownership returns to you.

Now consider another frontier of generative AI: text-to-image models. If you haven't tried them, stop reading and give them a go. Midjourney and OpenAI's DALL·E are my top recommendations. You type a prompt, and the system generates an image. The results are often stunning. The technology is breathtaking – creating photorealistic visuals or stylized artwork in seconds.

But for all their beauty, their limitations are clear. These models don't yet offer the kind of interaction that makes text-based interactions with ChatGPT so powerful. And in that sense, their shortcomings mirror a broader limitation in generative AI today.

Take Midjourney. I once asked it to create an image of a woman with flowers in her hair. It gave me several versions. I picked one and asked for a sideways-facing version. The model obliged – but her face had changed, and the roses were now lilies. I revised the prompt, started over, and nearly an hour later, got something closer to what I wanted – if I was willing to compromise.

That's the problem. These tools are built to create, to bask in the glow of their brilliance, not to collaborate. They don't handle nuanced feedback. You can't say, "Keep everything the same, just rotate her head." They regenerate everything. Iteration isn't built in.

But we're starting to see progress. Midjourney now offers inpainting, allowing regeneration of specific areas. OpenAI is improving how DALL·E responds to prompt refinements. These are steps in the right direction – but we're not there yet. Interactivity can't be bolted on. It has to be foundational.

And this isn't just about art. It applies to any domain where AI meets human creativity. A truly generative system should co-create – taking cues from us, adapting, listening, and leaving room for expression. It should help, not hinder. If you know what you want, you should be able to shape it freely.

Because the real promise of AI isn't in dazzling us – it's in partnering with us. Until that becomes the norm, we'll keep returning to the same truth: the future belongs not to the systems that perform *for* us, but to the ones that work *with* us.

Before we bring this chapter to a close, let's turn to one final application of creative machines, one that sits well outside how we typically think about these remarkable systems. And yet, when our creativity is supported, our humanity can grow – even in life's most difficult moments.

Mourning and the Machine

Before a child speaks, it sings.
Before they write, they paint.
As soon as they stand, they dance.
Art is the basis of human expression.

– Phylicia Rashad

Los Gatos, California
July 2024

It's a lonely Sunday afternoon. My husband and son are visiting my in-laws, and I've stayed behind to look after our fluffy little dog, Pucci (short for Puccini), and to prepare for an upcoming conference. Stubbornly, I go about our usual weekend routine – a long walk downtown, followed by a steaming cup of coffee. Yummy.

But it isn't the same. The bright California sun offers no comfort, the brisk walk no delight. I tell myself it's because my husband, David, isn't here. That must be it.

Back home, I glance at my phone. Five missed calls from my dad.

Before I have a chance to think – before I let myself think – I call him back.

"You need to come home now. The doctors say she only has a few days left."

I usually try to stay composed when talking with my dad. He's had it harder than any one of us since my mother's cancer diagnosis. But this time, I can't help myself. Tears flood my face, my voice trembles.

"I'll be there as soon as I can."

I take the redeye from San Francisco to Toronto, hop into an Uber, and race to the city's top cancer facility.

I see my mother in the early hours of the morning.

She passed three days later.

My mother, who looked just like me, lost her battle to cancer only months after her diagnosis. I see her face every time I look in the mirror. Truth be told, this entire book has been written in the shadow of my mother's passing. I still cannot believe that I will never see her face, never hear her voice again. She will never see this book published, I won't get to give her a copy.

In the weeks following her passing, I learned that grief is a quiet, secret club. A hidden basement in the house of our lives. No one talks about it – until I bring it up. Then the stories pour out. Everyone has one.

Grief is, for better or worse, universal. But what does any of this have to do with creative machines?

■ ■ ■

In 2019, Alison Pease stepped onto a stage in Charlotte, North Carolina. I've admired her work for years. There's a calm strength to her, a gentle gravitas. Her large blue eyes are both kind and sharp.

She begins her presentation, and I can already hardly sit still. She's talking about creative machines for mental health. My heart races. I've been exploring the same ideas, but I've never seen anyone else approach it quite like this.

A year later, Alison, her PhD student Lee Cheatley, Professor Wendy Moncur, and I conducted a study on how ALYSIA, the song-writing AI developed by my company, WaveAI, can help people navigate grief (Cheatley et al. 2022). Lee leads in-depth interviews, guiding participants as they write songs about their lost loved ones.

Most are not musicians. At first, they hesitate, unsure of where to begin. But ALYSIA eases them in. The machine suggests lyrics one line at a time, never imposing, only offering. They choose what resonates, weaving poetry phrase by phrase, gaining confidence as the creative process unfolds. Notably, just as my brother had discovered that ALYSIA's lyric generator was the most helpful aspect of the app, the study echoed this insight, emphasizing the usefulness of this particular component.

"Some of the phrases it suggested were really good – some even funny. In the end, I used my own lyrics, but I don't think I would have gotten there without it," one participant shared.

From time to time, the AI's suggestions strike an unexpected chord. Participants uncover emotions they hadn't fully recognized. AI helps them access buried feelings. That's the secret, if there is one – by listening to us, and helping us explore creative possibilities, machines can help us engage more deeply with our emotions.

A machine cannot grieve, but it can sit beside us as we do. It offers no judgment, no expectations. In its presence, there is no fear of saying the wrong thing or being misunderstood. And in that absence of judgment, people feel safe – free to express their grief honestly, to create without hesitation, and to confront emotions they might otherwise keep hidden. Sometimes, this safe, judgment-free space is exactly what's needed to access the scariest corridors of our psyche.

Self-expression is essential to healing. Research shows that creative expression, particularly songwriting, helps people process grief (Love 2007). By writing with the AI, participants took meaningful steps toward acceptance, self-discovery, and connection. They reminisced, redefined their loss, and, most importantly, remained in control of their creative process.

Beyond the study, I had witnessed the same tendencies among LyricStudio users. People turned to songwriting to process bullying, heartbreak, and deep personal loss – sharing truths in lyrics that might remain unspoken in everyday language. For me as well, creating music alongside our AI systems has been a powerful way to navigate grief and emotional healing.

This is the essence of humble creative machines: they offer a mirror, a space to explore our inner world. They extend a patient, gentle invitation toward self-understanding. By helping us reconnect with our creativity, they can also help us process our emotions – even in life's most difficult moments. I can think of no application of machines that touches our humanity more deeply than one that helps us feel, and walk through, the challenging but deeply human terrain of loss and bereavement.

As I am writing this, it's been nine months since I lost my mother. The grief process has been brutal. Whenever I need to feel her presence, I sit by my piano and play the song I wrote many months ago, in collaboration with my lyrical AI. Whenever I play it, or even just read the lines, tears pour out of me, as if on command. It holds everything I feel for my mother – the love, the loss, the tangled web of complex emotion. The push and pull of our connection when she was here, and the unsettling weight of her absence. Prose could never hold all of that. But poetry? Poetry can.

Here is an excerpt. Mom, this one's for you.

The Wild One

In spite all of your strife
You lit in me the flame of life
You gave me big blue piercing eyes
And when I look up at the skies
I feel your love run down my spine
I am your child
The wild one

The wild one
You hear music in the wind
The wild one
You're dancing with the trees
The wild one
Who lives beyond the veil

12

A Fresh Look at Familiar Topics

So far, I have offered you a new lens on creative machines, arguing they reflect humanity in all of its glory and folly and that it would be best for them to play a humble role in our world. This provides a foundation for how to move through this perplexing new age of AI. In this chapter, we are going to turn toward some hot topics and look at them with fresh eyes.

My goal over the following pages is to challenge you to consider some new perspectives on popular topics. Many of these issues have been explored extensively in academia well before the proliferation of creative machines. But everyone, including academics, has evolved their views on these critical topics once their impact on the world became apparent. So what I will share with you here integrates old academic ideas as well as criticism of common discourse. But mostly, I will offer my own personal take on the issues, informed by a combination of my experience in academia and the startup ecosystem.

Let's address some pressing and practical issues on how to create a better world given the constraints of our well-established systems.

Job Displacement

"It'll be fine," they assure us. "Don't worry. Even if you lose your job, something better will come along soon enough." This optimistic narrative, rooted in historical precedents, suggests that technological advancements always lead to better jobs and brighter futures. That there is some type of universal law that ties progress in technology with progress for all of humanity. After all, history is filled with examples where innovation created new opportunities, improving lives and industries alike. But this rosy narrative doesn't hold up to scrutiny.

Consider the plight of horses. For many years, horses toiled in agriculture, often harnessed to heavy equipment or walking in endless circles to power mills. With technological advancements, their roles evolved – they began pulling elegant carriages through bustling cities. It seemed like progress had improved their lives. However, with the advent of the automobile and mechanized farming equipment, millions of horses were rendered obsolete. Many were sent to slaughter, victims of a technological shift that had no place for them. The grim lesson is all too relevant: just because progress improved conditions in the past, it doesn't mean it will continue to do so. Technological advancement does not guarantee better jobs or outcomes – not for horses, and not for humans.

The Industrial Revolution provides a cautionary tale. It transformed the production of goods, making them faster and cheaper than ever before. The resulting economic growth and emergence of the middle class laid the foundation for modern life.

Yet, the early years were marked by immense suffering. Workers faced long hours, unsafe conditions, and the loss of communal village life. Child labor was rampant, and the transition left many displaced. These "growing pains" were anything but minor; they came at a profound human cost. The parallels to today are clear. Generative AI is poised to bring tremendous advancements, but the transition period may be fraught with challenges. Job losses, the need for re-education, and economic upheaval are serious issues that require immediate attention. Blind optimism – the assumption that "it will all work out" – is dangerous and dismissive of the very real pain people may endure.

As we enter the age of generative AI, we can feel the landscape of work shifting beneath our feet. Entire industries are being redefined, and the ripple effects are already visible. Software development, one of the most lucrative fields, is experiencing significant changes. Generative AI tools, such as GitHub Copilot, enable developers to write code faster and more efficiently. If one developer can now do the work of three or ten, layoffs are inevitable.

Marketing is undergoing a similar transformation. Early applications of OpenAI's models, like Jasper, demonstrated how AI could revolutionize copywriting and content creation, leading to billion-dollar valuations before the rise of ChatGPT. Now, with greatly improved tools like ChatGPT widely available, as well as high-quality text-to-image and text-to-video models such as Midjourney and Sora, the demand for human marketers is reducing, and additional layoffs may be difficult to avoid.

Many cling to the belief that certain jobs – those requiring creativity or emotional intelligence – are safe from AI. But as this book has explored, machine creativity is real, and its distinctions from human creativity do not make it less valuable or less of a threat. Likewise, emotional intelligence may not be the insurmountable barrier it appears to be. Machines don't need to replicate our unique creative processes or emotions to pose a significant employment threat across a wide range of industries. The comforting idea that machines will never compare to humans risks blinding us to the challenges already unfolding.

Creative industries hold a special place in human culture. Musicians, artists, and writers derive immense satisfaction from their deeply personal and expressive work. Their creations enrich lives, allowing others to connect with universal themes through the unique and beautiful lens of an artist's soul. Yet, generative AI threatens to encroach upon these sacred spaces. While there is hope that AI can serve as a co-creative partner – elevating human artistry, as I've outlined in Chapter 11 – it is unlikely to be the only path pursued, as others prioritize using AI to replace rather than enhance human creativity.

Similarly, professions rooted in emotional intelligence, such as therapy and counseling, are facing new challenges. AI-powered therapists, for example, already offer clients a level of privacy and non-judgmental interaction that human therapists cannot replicate. Relationship-centered roles, such as sales, are also vulnerable as AI systems grow increasingly sophisticated in understanding and responding to human emotions. The rise of these capabilities reveals how even deeply human domains are no longer safe from the reach of machines.

Even if we imagine an optimistic future where generative AI creates better, more fulfilling jobs, the short-term impact cannot be ignored. Families will face financial instability, and workers will require costly and time-consuming retraining. Worse yet, it remains unclear which careers will be worth retraining for. The burden of this transition will fall disproportionately on those living through it. Policymakers must act decisively to mitigate these effects by investing in robust social safety nets and exploring solutions to guide workers through this uncertain shift.

The broader implications are even more sobering. Over time, as generative AI reaches its full potential, society will need to grapple with a world where much of today's labor is unnecessary. Universal basic income could become essential to address unprecedented levels of unemployment and inequality. Moreover, if enough people are displaced from the workforce, there may no longer be sufficient consumers to sustain demand for goods and services, ultimately jeopardizing the profitability of even the largest firms. Taken to its logical conclusion, this dynamic could destabilize our entire economic system.

The challenges posed by generative AI are immense, but so are the opportunities. We must approach this moment with creativity and resolve. Policymakers, researchers, and industry leaders must work together to design systems that prioritize human well-being. On an individual level, the best defense is adaptation. Learning to use AI tools effectively and staying at the forefront of your field can provide some measure of job security. However, systemic change is essential to ensure a future where technology serves humanity rather than leaving it behind.

This is a crucial moment in history. Generative AI has the potential to create a future where people work less, live more, and spend more time with their families. However, this vision won't materialize on its own. We must take deliberate action to shape the future that ourselves and our children would want to live in.

Other than job displacement, few questions loom larger than those around the closely related questions surrounding data rights and intellectual property. This is one of the fiercest debates of our time – touching on ownership, consent, and the future of creative work. It's a complex, tangled knot of ethics, economics, and power. Let's begin to unravel it.

Data, Identity, and IP

The use of data in AI systems – and the intellectual property storm swirling around it – is one of the most urgent debates in generative AI. Unfortunately, the conversation has been flattened into extremes. So how do we actually resolve this? Are there practical ways forward that protect artists – options companies could realistically implement?

The terrain is messy and full of competing interests. But let's start with a simple, albeit uncomfortable, truth: tech companies have been profiting from our data for years. Personalized feeds, recommendations, targeted ads – these were all built on behavioral data collected en masse.[1] You're not just the product; you're part of the training data. Long before anyone typed "in the style of Studio Ghibli" into an AI prompt, using data without permission or compensation was already powering the world's most valuable companies.

So when firms like OpenAI trained their models on vast amounts of publicly available data, it wasn't a novel concept. What changed was the application. These models didn't just use data to improve user experience – they enabled, and in some cases openly encouraged, imitation. This is a very specific and deeply problematic use: training on artists' data in order to generate work in their distinctive styles.

Early demos didn't just show off capabilities – they explicitly suggested users enter the names of living artists and photographers. The results mimicked their styles with eerie precision. This led to their aesthetic work being drowned in algorithmic

---

[1] One example of how our data has long been used freely is collaborative filtering, or clustering – techniques that group large numbers of people into "prototypes." Simply put, if people like you enjoy a piece of content, you're likely to enjoy it, too. This method underpins many successful products that recommend your next favorite playlist or sitcom.

imitation. The creative labor of real people, repurposed to mass-produce imitations without consent or compensation.

I doubt companies thought what they were doing was ethically sound. But from a business standpoint, it makes a certain kind of brutal sense. Established artists already have a style, a name, a following – so what better way to showcase a new technology than by leveraging that recognition? It worked. But the cost was enormous: not just to the artists themselves, but also to the broader legitimacy of AI. It made generative models look like machines of imitation, not instruments of creativity – a perspective that many still hold today.

Artists pushed back. But it's important to note that the early objections weren't just "don't imitate us" – they were "don't use our data." And understandably so. However, the type of generative AI we see today *requires* massive datasets. But it does *not* require imitation. These are separate issues, and conflating them limits our ability to respond.

Copying and learning are not the same. Humans have always drawn inspiration from one another without descending into plagiarism. We expect people to walk the line between influence and theft. Machines can be designed to do the same – but only if that's a deliberate goal. Learning without imitation is technically feasible. It's not always standard practice, but it's well within reach – and some companies are already moving in that direction. Just not far enough.

AI companies shouldn't just stop imitating specific living artists and encouraging users to do the same – they should make it *impossible*. This isn't a fantasy. Safeguards against imitation can be built in at multiple stages, from training architecture to prompt filtering. This is a solvable, technically feasible problem.

Still, one might ask: why compromise at all? Why not demand that AI companies both *stop imitating* and *fairly compensate every creator* whose data was used? It's a reasonable instinct – but this is where we run into one of the most overlooked and unfortunate realities of this field.

The truth is that if every creator were properly compensated, the economics of large-scale generative AI would collapse. The reason these models work is because they're trained on massive datasets – tens of millions of examples, often more. If each piece of content cost even a hundred dollars (hardly a great value for a piece of art or music), the bill would reach hundreds of millions, possibly billions, before even accounting for the model training, staffing, and operations. Only the richest companies or countries could afford to play. That's a direct path to monopoly – and it raises the uncomfortable possibility that by fighting for fair compensation, we may unintentionally strengthen the gatekeepers. There are other complexities tied to strict data laws – chief among them the risk of surrendering AI leadership to countries that deliberately maintain looser regulations.

More interesting, though, is a quieter reality: some AI companies *are* paying for data – just not to the creators. Since it's economically infeasible to compensate every individual in a massive dataset, the money often goes to whoever owns large blocks of content – music labels, stock image companies, rights aggregators. These are often well-established, well-funded entities. The creators whose work fills those datasets don't see a dime. So when we call for compensation for our data, what often happens in practice is that one type of company pays another – while the artists and musicians who actually made the work remain entirely excluded. It's a redistribution of wealth that neatly bypasses the people who created the value in the first place.

But data use alone isn't the only threat. Sometimes, AI can imitate us *without* needing our original work at all. Take OpenAI's recent voice assistant, whose voice sounded like Scarlett Johansson (Tassi 2024). According to reports, the company approached her to use her voice. She declined. They used a sound-alike instead. Her actual data wasn't used, but her likeness was clearly evoked.[2] This matters: a system can violate someone's identity without ever touching their files.

That's why part of the conversation must shift from how companies build AI to *what they build it for*. In the human world, we don't regulate how people think – we regulate how they act. Society doesn't legislate what goes on in your head, but it does care about what you do. There's wisdom in applying this principle to creative machines.

Ultimately, we need to ask the most important question: *What does the AI actually do with what it's been trained on?* That remains critical even if imitation is avoided entirely. Because for many artists, knowing their work helped build a system that could eventually replace them is no less unsettling. They may not be copied – but their labor is still fueling their own displacement.

And the risk doesn't go away. These systems can be used to automate creative work at scale – not by plagiarizing, but by replacing. A system doesn't have to imitate to disrupt.

This is why the core issue isn't just data – it's outcome. That's why people didn't march in the streets when companies used their data to fine-tune recommendation algorithms. But this is different. The impact is different. The stakes are higher.

We must begin to disentangle these layers: data use, imitation, and economic impact. They're connected, but not the same. Right now, too much of the conversation collapses them into one vague grievance – making it harder to craft effective responses, and often leaving artists with the short end of the stick.

We're unlikely to stop the trajectory entirely. Some investors will always back founders with the goal to replace creatives. That's why I believe in the critical role of *humble creative machines*:

---

[2] It's worth noting that this story did eventually resolve with OpenAI removing the sound-alike voice from their platform (Tassi 2024).

systems built not to replace human creativity, but to elevate it, which we discussed in Chapters 10 and 11. These tools won't stop the full automation crowd. But they offer a counterweight. A human artist, empowered by a humble creative machine, can outperform an autonomous system.

The uncomfortable reality is that even well-intentioned efforts to protect artists can sometimes end up reinforcing the very systems that exclude them. Calls for compensation often funnel money to intermediaries, while the creators themselves remain sidelined. Imitation may be the most visible harm, but it's not the only one. What ultimately matters isn't just how these systems are built – it's what they actually do. The ideas in this chapter are not meant to offer final answers, but to help untangle a complex landscape, in the hope of building a world where human creativity not only survives but continues to grow and thrive.

When machines are trained on our words and works, the threat isn't just theft – it's also distortion. Because once the data is ingested and the output begins, it doesn't just imitate us, it influences what we believe. Let's follow this path into the complex terrain of truth itself, where AI doesn't just echo us, it edits reality.

Truth and Misinformation

Truth. Lies. Perception. Deception. AI doesn't just reflect our relationship with reality – it warps it, sharpens it, makes it harder to see where fact ends and fiction begins. If misinformation is a spark, AI is the wind, fanning it into a wildfire of certainty, confidence, and confusion.

Misinformation has plagued humanity since the beginning of time. Before the written word, before literacy, knowledge was preserved through stories, carefully passed down across generations. But power shaped the narratives. Monarchs, religious institutions, and rulers controlled access to information, dictating reality to those who had no way to verify it. Then came books, and with them, the promise of independent truth-seeking. But the burden was immense – long hours in dusty libraries, sifting through conflicting sources, searching for reliable facts. Information was still a privilege, reserved for those with the resources, time, and education to seek it out.

Then came the internet, and suddenly, information was at our fingertips. It should have been the dawn of a new era – one where access to knowledge empowered people like never before. But that's not what happened. Instead, we found ourselves drowning in an ocean of information, unable to tell what was real and what was fabricated. Social media, in its pursuit of engagement and profit, sealed the deal. Echo chambers formed, filtering reality through algorithmic precision, ensuring that each person saw only what reinforced their existing beliefs.

Polarization followed. Now, entire segments of society believe that everything their political opponents do is wrong, laughable, and corrupt beyond redemption. No nuance. No curiosity. No loyalty to truth. Because in a polarized system, admitting the "other side" has a point makes you a traitor. Truth isn't the goal – winning is. We see this across every contentious issue: immigration, healthcare, abortion, you name it. The line has been drawn. There is a right side and a wrong side, and where you stand depends on your tribe. The old tradition of simply reporting facts – already challenging – has gone out of fashion. Instead, we get performances of outrage, shouting matches between opposing echo chambers, each side more concerned with their narrative than reality itself.

And then came AI.

AI didn't invent misinformation. It simply amplified it. We already lived in a world where anyone could post just about anything, unchecked and unverified. Now, with AI tools in hand, creating misinformation is easier, faster, and more convincing. Text-to-image models can fabricate compelling visuals of politicians doing whatever you want them to be doing – petting kittens or saluting beneath a communist flag. Large language models confidently generate answers even when they don't know, without any hint that the information they are providing is fabricated. And because AI responses feel authoritative, because they come neatly packaged in perfect grammar and confident phrasing, we are tempted to believe them.

But the truth is not always certain. And that is uncomfortable. Science, the pursuit of truth, has always required curiosity and flexibility. The entire scientific community has, time and again, revised its understanding – sometimes discarding once-accepted facts and reviving ideas previously thought obsolete. Even artificial neural networks were once dismissed as ineffective, only to become the foundation of modern AI. Truth is a shifting landscape, and real truth-seekers must hold uncertainty in mind, knowing that what we accept today may change tomorrow. But in a world that craves certainty, that discomfort is too often avoided.

The problem isn't just AI – it's us. It's the way we interact with information, the way we prioritize comfort over truth. We like the stories that confirm what we already believe. We like the validation. We like the feeling of being right. Real truth-seeking is difficult, uncomfortable, and time-consuming. If you really want to understand a complex issue, you need to read dozens of articles, verify sources, dig into peer-reviewed studies, cross-check historical records, engage with experts who challenge your views. And even then, you may find yourself lost in uncertainty, forced to accept that truth is complicated, that it resists simplicity, that it doesn't always align itself neatly with your feelings or the beliefs held by your tribe.

We say we want the truth, but do we? More often than not, what we really want is reassurance. We want to feel validated, to belong, to avoid the cognitive dissonance of realizing we might be wrong. The irony is that we deceive ourselves just as much as we deceive others. Studies show that most people rate themselves as better-than-average drivers – statistically impossible. We tend to overestimate our intelligence, our attractiveness, our abilities. It's a universal human trait, a subtle self-deception that smooths over reality.

Having lived across three continents before the age of thirteen, I am struck by how differently reality is understood from place to place. Even historical facts – like World War II – are taught in dramatically different ways depending on the region. My point isn't that people are lying, but that through different lenses, different priorities, and different selections of events, the same period can be painted in entirely different ways. The differences are sometimes so stark that the same era can feel almost unrecognizable.

AI makes this problem worse, not necessarily because it is malicious, but because it is designed to give answers, not to hesitate. Truth is messy, uncertain, full of challenges. AI, by contrast, is optimized for clarity and coherence. And because it is created by the world's largest corporations, it is ultimately aligned not with truth but with the interests of its creators – at least when the truth conflicts with those interests. These companies have fiduciary responsibilities; their goal is not to inform you but to maximize shareholder value.

Do you really think that objective, unfiltered truth is their priority? We've already seen how social media companies profit by building and reinforcing echo chambers. Do we really believe that AI companies will act more ethically? And when AI models must align with the priorities of foreign governments, the "truth" they provide may diverge even further from what we see as reality. Sometimes a position can be overt, but often the perspective is buried in subtle phrasing and fact choices that can be easily designed to steer public opinion. When every major AI model presents a uniform narrative on key issues, it does not mean that narrative is correct – it only means that control over information has been consolidated.

So what do we do? Some believe the solution is to make AI more truthful, to align it better, to refine its training so that it stops lying to us. But that's the wrong battle. The real danger isn't just AI's misinformation – it's our willingness to accept easy answers. AI should never be allowed to be the arbiter of truth. Not because it is evil, and not just because of incentive misalignment, but because truth is too important, too complex, too human to outsource. The moment we surrender our responsibility to seek it, we hand over the keys to reality itself.

Philosopher Shannon Vallor (2024) warns of our "evolving and troubled relationship with the machines we have built as mirrors to tell us who we are, when we ourselves don't know." She highlights the danger of surrendering our agency, seduced by the illusion that AI understands us more clearly than we understand ourselves.

Truth, knowledge, and the struggle for self-understanding are complex endeavors – and the beauty lies in the struggle itself. When we remove ourselves from the ongoing conversation through which we make sense of reality, we don't just lose agency; we abandon one of the most essential aspects of being human.

These are not things an algorithm can calculate or a machine can retrieve. Truth is elusive, messy, uncomfortable. It demands patience, effort, and the courage to change our minds.

AI should serve as a wake-up call – a flashing warning sign of how easily truth, and our relationship to reality itself, can slip away when we value convenience over rigor. *When we outsource the struggle for understanding, we risk losing far more than just good information – we risk losing ourselves.*

There are no shortcuts. There never have been. From ancient oral traditions to dusty libraries to the bottomless feeds of today, truth has always required work. And now, in an era where misinformation moves at the speed of light, it's up to us to decide whether we're still willing to do that work – or if we'll let machines think for us.

The relationship between truth and fiction has always been more tangled than it seems. It's not just that fiction can be mistaken for truth – it's that, sometimes, it can become truth. The stories we tell shape the futures we pursue, especially when it comes to AI. Science fiction hasn't just predicted our trajectory – it's helped script it. But perhaps it is about time that we widen the frame.

Stranger Than (Science) Fiction

Before AI could make art, the arts have been inventing AI. The stories we tell shape the technologies we build. For decades, science fiction has not just imagined AI but has also crystallized our expectations of it. The visions of AI in books, films, and television have bled into reality, influencing the very entrepreneurs, programmers, and investors who bring AI to life.

In a stunning twist, the narratives that once opened our imaginations now pose a major hindrance. They create a distorted lens through which we perceive AI, leading us to fear it, misunderstand it, and limit its possibilities. For all of our love for science fiction, we need to look at it critically – to break free from the narrow perspectives that no longer serve us.

Science fiction presents AI through an astonishingly narrow lens: as human-like, as rapidly and almost magically evolving, and as a force of existential danger. Anthropocentrism dominates AI in science fiction. This false narrative blinds us to a greater truth: intelligence is not a singular, linear spectrum with humans as the gold standard. AI does not need to mirror us to be valuable. Yet, science fiction refuses to imagine it as anything else. From *Her* to *Ex Machina* to *Westworld* to the striking twist in *Lifelike*, AI is almost always framed in human terms. It looks like us, sounds like us, and thinks like us – until, inevitably, it either falls in love with us or decides to annihilate us. AI is either our servant or our superior, but it is always defined in relation to humanity. If it is less intelligent than us, it is a mere tool. If it surpasses us, it is a crisis.

Even in stories where AI is benevolent, such as *Star Trek*'s Data or *Bicentennial Man*, it is still always in pursuit of one thing: becoming human. The idea that an advanced intelligence might exist without the slightest desire to mirror us is almost unthinkable in science fiction. But why? Why is our imagination so limited that we cannot conceive of AI existing as something wholly distinct – something that neither loves nor loathes us, but simply is?

Then comes magical emergence – this idea that intelligence, once sufficiently advanced, must wake up. This assumption leads directly to one of science fiction's most gripping yet misleading tropes: the idea that AI, once it achieves a certain level of intelligence, will inevitably develop emotions, goals, and self-awareness. From HAL 9000 in *2001: A Space Odyssey* to the machines in *I, Robot*, science fiction repeatedly assumes that intelligence alone begets intent. Yet, in reality, AI does not desire. It does not fear. It does not plot. It is humans, flesh and blood, who apply AI to various ends, some benevolent, some selfishly motivated by short-term gains.

This fear-driven narrative makes for compelling storytelling, but it is completely divorced from reality. The real danger is not AI gaining consciousness and waging war on humanity – it is how we, the ones with actual consciousness and intent, choose to design and deploy it. The danger is that stakeholders will choose to utilize AI toward selfish ends, with little regard to employability and human wellness, furthering the gap between the haves and have nots. AI already surpasses human capability in many areas, from language translation to medical imaging to large-scale prediction. But none of these advancements imply self-awareness, let alone intent. Intelligence does not automatically give rise to desire, ambition, or malice.

This matters not just as a critique of a genre, but as a lens through which we shape reality. If we continue viewing AI through the same narrow lens, we will fail to realize the true opportunities and the real risks before us. Even investors and entrepreneurs are deeply influenced by science fiction narratives, understanding risks and opportunities through a rather narrow lens. Consumers fear or embrace AI based on these ingrained ideas.

But what if science fiction offered a different vision – one that not only reshaped our understanding of AI but also influenced how we build and interact with it? What if these stories encouraged AI development that fosters creativity, collaboration, and human potential, rather than reinforcing outdated fears?

What if, instead of showing AI as a power-hungry, human-like being, it envisioned AI as something profoundly beautiful – something that humanity uses to elevate itself? Imagine an AI akin to Yoda, a humble, deeply insightful machine that doesn't merely offer knowledge but actively nurtures creativity, helping a young woman on her quest to find her voice and become a musical superstar. Unlike traditional AI portrayals, this machine would not seek control or independence.

Or, imagine an AI not as a singular entity, but as a vast, interconnected collective mind – an ever-growing reservoir of knowledge and creativity that anyone can tap into, expanding their intellectual and artistic pursuits. What if AI was not a looming adversary or a subservient machine, but an extension of human potential – a tool that makes us greater than we could ever be alone? What if science fiction dared to break from its familiar tropes and take us somewhere entirely new? Oh, the places we could go.

We have barely begun to explore what AI can become beyond the narrow confines of science fiction's imagination. While there are exceptions – such as *The Culture* series by Iain M. Banks, which envisions AI as benevolent and symbiotic with humanity, or *Person of Interest*, which explores AI as a nuanced and morally complex entity – they remain rare. It is time to reclaim our own creativity. To break free from these outdated myths. To see AI not as a narrow reflection of ourselves, but as a new and uncharted frontier.

We must reimagine AI – not as a child striving toward humanity, nor as a god towering over it, but as something else entirely. Something brilliant. Something inspiring. And in the meantime, as we wait for science fiction to evolve, we can choose to break free from these limiting narratives. We can see AI for what it truly is, unshackled from the constraints of familiar myths. AI does not have to strive for humanity, nor does it have to be our undoing. It can be something entirely new. Something that, ironically, only the boldest of science fiction writers might dare to dream.

Yet, even if we break free from the narrow lens of science fiction, I believe there are realms so profoundly human that machines, no matter how advanced, should never intrude upon them. So let's shift our focus and explore the boundaries that AI should never cross.

13

Embracing AI's Limits

"Nothing is lacking, and yet we search."

– Lao Tzu

> *Fire won't chase away the night,*
> *No tide can cleanse our weary plight.*
> *We are not riddles to be solved,*
> *But dancing verses yet unknown.*
> *Not all that flows must seek its key,*
> *Nature plays her symphony.*

We are always searching for the cure to our unease, for the ultimate key to happiness. Throughout history, we've chased myths and miracles: the Philosopher's Stone, the Fountain of Youth, the great machines of the Industrial Revolution. Each promised ease, liberation, transcendence.

And now, as we enter a new revolution, one powered by creative machines, the need for wisdom and balance has never been greater.

Petal Portraits

She was breathtaking – flowers blooming from her hair, vines embracing her face. She was Earth itself. Petals of every shade spilled from her greenery, wild and untamed. The trees, the flowers, the fragrance of it all. Mother Nature.

That vision changed me. It reminded me of what matters.

Now my little white kitchen table is cluttered with dried flowers, a pint of glue, a repurposed makeup brush, and stacks of tiny canvases. I've been working with AI-generated images of women – rendered in Midjourney – redrawing them in pencil and charcoal, layering something unmistakably human over something born of code, and intertwining it all with nature.

When the image feels right – delicate, grounded, alive – I press it between glass and seal it into a golden frame. It's not exactly what I envisioned, but it's beautiful in its own imperfect way. My house is running out of space. It's crowded with these floral portraits – images of women surrounded by pressed flowers.

At first, I didn't understand why I was so obsessed with re-creating her. The Earth-mother. The goddess of vines and petals. I kept trying – over and over, switching materials, playing with composition – chasing something just out of reach.

When I painted in my teens, it was different. Paint on canvas, sometimes a pencil sketch. Today, my art demands raw materials. I collect dried bark. I press wildflowers. I want to feel nature beneath my fingertips – though I live among redwoods, I still can't seem to get enough of it under my skin.

The human-made parts matter, too. The glass. The glue. The silk and lace I sometimes add. The golden metal frames. And, at the very center, always, is an AI-generated image – iterated end-lessly in Midjourney until I land on just the right silhouette.

It's strange, isn't it? Why am I working so hard to recreate this vision? Sometimes I look at what Midjourney spits out – hauntingly perfect images conjured in seconds – and I wonder: why bother making anything at all? Why spend hours on what a bot can do in milliseconds?

And the answer, for me, is this: because it's not the same. Because what I'm making tells a story. My story. Our story.

When I slow down enough to feel the chaos in my chest – the world's turmoil mirrored within me – I hear the silent scream of Mother Earth. The story of creative machines must be woven back to the source, reconnected to her vines, and reshaped to serve the living, organic world that is the heart of all creation.

So I give you this unplanned chapter. It emerged from me like a branch from a tree. My small acts of creation – messy, handmade, slow – felt like a quiet resistance. I can't shake the feeling that we're tilting too far, too fast – and if we don't pause, we might just sacrifice the most wondrous parts of being human.

Lest We Become Machines

Before the machine, before the factory, the sun called us to rise, the stars signaled rest. A farmer tended the land, then set down his tools to share a meal, to laugh, to gather in the warmth of his community. The rhythm of life moved in harmony with the earth, replenished in the flickering glow of firelight in cer-emonies where voices rose in song and feet met the ground in dance.

Then came the factory. The Industrial Revolution turned time into a commodity, dividing life into shifts and schedules. The rhythm of the sun no longer dictated the day – the clock did.

The bell rang, and work began. The bell rang, and work stopped. But only for a moment. The machine never truly rested, and neither could the workers who fed it.

That mindset never left us. It just evolved.

Now, we are at the dawn of a new revolution, one shaped by generative AI, with machines that can think and create. AI promises to free us from tedious work, yet instead of using it to reclaim our time, we risk becoming even more machine-like ourselves.

Behind every AI system, hidden from view, is an army of human workers performing the digital equivalent of factory labor. Annotation workers in Southeast Asia and Africa spend their days labeling images, refining chatbot responses, and performing repetitive tasks that teach AI how to please Western users. They are paid little, given no control or creative freedom, and treated as disposable.[1]

This mechanization of human beings isn't limited to low-paid labor. A C-suite executive sits at her desk late at night; she is illuminated by the blue glow of her monitor. Her inbox is full, Slack notifications still ping. She fights her exhausted body to respond. Her child is asleep in the next room. She missed bedtime again. She tells herself she will make up for it later – on the weekend, next month, when work slows down.

Even those who have reached the top find themselves trapped. They pay exorbitant mortgages for homes they barely have time to enjoy. The system ensures that no matter how much they earn, they are always reaching, always working, never fully free.

The pressure to perform, optimize, and produce isn't just exhausting – it's making us sick. Burnout, anxiety, and depression are soaring (Osorio and Hyde 2021; Smith 2023a; Witters 2023). Our culture treats rest as weakness and self-worth as something

---

[1] OpenAI's outsourced workforce in Kenya, earning as little as $2 an hour, was tasked with filtering through horrific content so that AI could appear safe for users (Dzieza 2023).

to be earned. We were not built for this. The human nervous system was never designed to run at the pace of an algorithm.

Surely, every creature must labor to survive, but none are caught in an endless cycle of exhaustion. A lion hunts because it must eat, but once it has fed, it lies in the sun, guilt-free, surrounded by its pride. A bird builds a nest, not for profit or endless optimization, but because that is what life calls for in that moment – and once the work is done, it returns to its flock, where safety and companionship are found.

The machine is not the enemy. The real threat is that those in power will use AI to extract more labor while giving back less. There is a growing risk that, instead of using AI to enhance human potential, we will be expected to compete with it – forced to work faster, produce more, and become more machine-like ourselves. In a world where humans compete with machines, the work conditions for humans may rapidly deteriorate.

Tech workers are already seeing the early signs. As AI automates coding, companies are not using these tools to empower employees but to justify layoffs and demand more output from those who remain. Perks are disappearing, work hours increasing, and the future is uncertain at best.

What about creative work? When we think of actors, perhaps Jennifer Lawrence comes to mind; for music, maybe Taylor Swift. The celebrities at the very top will be fine. But the vast majority of creatives are not perched at the summit of the cultural pyramid, and they are vulnerable. Writers, artists, and musicians fear that their fields may soon be overtaken by AI, leaving human creatives struggling to compete against tireless machines.

The 2024 Screen Actors Guild strike exposed this trend in entertainment (SAG-AFTRA 2024), where studios attempted to secure the right to scan background actors and reuse their digital likenesses indefinitely – hinting at a future where performers are no longer needed at all. If corporations are already pushing to replace workers before AI is fully capable, what happens as these technologies improve?

We must realign our vision. We are not here to serve the machine. The machine is here to serve us.

If we are not careful, we may find ourselves forced to work like machines, even more so than we do today – not because AI demands it, but because capitalism does. But this is not inevitable. We must resist the narrative that productivity is the highest measure of worth, that efficiency justifies human suffering. And most importantly, we must refuse to let ourselves be reshaped in the machine's image.

Beneath all of this – beneath the schedules, the algorithms, the constant push to optimize – something within us resists. A force that cannot be mechanized, cannot be controlled.

A knowing.

An instinct.

The Instinct to Be Human

Everything shifts,
Yet it's all the same.
Fog dims the light,
But can't snuff the flame.

Ancient and free,
Etched in our bones,
A primal beat,
Beckons us home.

A zebra lies motionless, frozen in terror, as a lion's teeth graze its flesh. Then, suddenly, the lion is forced to flee. Waking from the grip of death itself, the zebra begins to shake. Tremors ripple through its body, shaking off the horror, bringing it back to life. When the shaking finishes of its own accord, it rises. The fear is gone. The zebra moves on, unburdened.[2]

This is the body's wisdom – a primal instinct to release trauma, to reset, to heal. All mammals have it. But we, with our complex minds and over-analyzing ways, do not trust it. When we are struck by grief, fear, or rage, our bodies want to shake, to purge the pain. Yet we resist. We freeze. We trap it inside, where it hardens into trauma and anxiety, festering over years.

We have forgotten how to be animals. We have forgotten how to trust ourselves.

And not just in pain. Somewhere along the way, we abandoned our ability to trust our own instincts across the board – our creativity, our passions, our ability to navigate the world without a constant flood of external guidance. We look to books, to experts, to algorithms to tell us how to eat, how to sleep, how to love, how to dream. We let the world dictate who we are, rather than feeling it in our bones.

Animals do not ask how to live. They just do. There was a time when humans, too, leaned into their own knowing. Ancient cultures did not see creativity, wisdom, or purpose as things to be handed down from authorities. Young people were sent on solitary journeys, not to be told who they were, but to discover it for themselves. Some Indigenous traditions still hold Ayahuasca ceremonies – not to instill the wisdom of a church or a guru, but because they trust that wisdom lives within each person, a direct, unbroken line to something divine.

Somewhere in the rush of civilization, in the noise of information, we stopped listening. We outsourced our instincts. We

---

[2] In his book, *Waking the Tiger: Healing Trauma*, Peter Levine (1997) describes how prey animals, after surviving a predator attack, instinctively discharge the traumatic experience through intense shaking, allowing them to recover without lasting trauma. In contrast, humans suppress such instinct, causing unresolved trauma to be stored in the body.

allowed a million tiny voices – on screens, in research papers, in social media – to drown out the voice within.

And now, with AI, we stand at a crossroads.

Let AI be a ladder to elevate us, not a replacement for our inner knowing. Our creativity and intellect are not to be extracted from us, commodified, and outsourced to a machine. It is something visceral, natural, something born from the body, deep within the magic of the animal world in which we occupy such a precious role.

And above all, we must not let AI – or any technology – replace our most fundamental human needs. A newborn needs their mother's love and contact; no machine can replace that, nor should it. Adults, too, need connection. We need our loved ones, in person, where human touch and presence shape our emotional well-being. Science has begun to validate what we've always known: touch calms the nervous system, strengthens bonds, and plays an important role in childhood development (Ellingsen et al. 2016; Cascio, Moore, and McGlone 2019).

I do not want digital friendships – I want my friends in my living room, or laughing with me in a café, sunlight on our skin. I don't want Zoom calls across the ocean. I want to sit across the table from my father, drinking tea, feeling the reality of his existence, his life intertwined with mine.

We are not brains in jars exchanging data one byte at a time. We are flesh and breath, instinct and feeling. We are wild things that learned to build cities, and the internet, and now creative machines. And no matter how far we advance, how intelligent our machines become, we must never forget the ancient wisdom written in our bones.

Let's shake off the illusion that AI will save us from being human. We don't need saving – we only need to remember who we are.

Our Greatest Creation

In our race to build artificial intelligence, we've forgotten something essential:

We already know how to create life.

And it's harder. Messier. Yet more sacred and valuable than anything made from code and silicon.

In my early twenties, I lived in my head. I was a classic academic overachiever – studying computer science, diving into theoretical foundations of machine learning, and preparing for a PhD. I believed intellect was the highest form of human expression.

Then I met David, the love of my life, and we decided to have a child.

I hadn't really thought about pregnancy. I assumed it would be… well, easy. Something the body just did, while I continued to do difficult work with my mind.

Before I had time to make sense of any of it, the first trimester hit like a tidal wave – nausea, exhaustion, disorientation. But there was something else, too. Something deeper. A force unlike anything I'd encountered in all my years of research. It overtook me and demanded everything – physical, emotional, spiritual.

I was creating life. And it was every bit as immense, and as difficult, as that phrase suggests.

I remember telling my mostly-male classmates: "Pregnancy is harder than research."

They stared at me in disbelief, speechless. But I still stand by it. Creating life is harder. And far more profound. You're not producing knowledge or a tool. You're producing *being*.

And yet, this act – the most human, most awe-inspiring act – gets less status and wonder than our machine-building. We talk about AI as if it were the pinnacle of creation. But what could possibly be more miraculous than bringing forth a conscious, living, breathing human being?

When I look at my son, I don't just see intelligence.

I see joy. Curiosity. Sorrow. Wonder.

I see the full, rich range of a human being.

No matter how much pride or affection I feel for the AI systems I've built – and I do – they will never compare. They will never hold the weight of a newborn in their arms. They will never look back at me with the eyes of someone who knows me before knowing what knowing is.

And anyone who has built intelligent machines and also held their child should understand this in their bones.

We are so quick to revere artificial intelligence.

We are willing to risk mass unemployment for AI.

We are willing to destroy ecosystems to train large language models.

We are willing to let go of the real in pursuit of the artificial.

It's time to remember what matters. AI can be powerful. Beautiful, even. It can enhance creativity, increase efficiency, open us up to new possibilities. It deserves our respect.

But it should never be confused with creation. Not the kind that grows fingers. And lungs. Not the kind that wakes you up in the middle of the night crying. Or laughing. Not the kind that holds your hand when you're old.

Let the machines do what they do. Let them help, support, extend.

But don't ask them to replace the greatest creation of all. Because they can't. And because they shouldn't.

14

Finale: The AI Symphony

We stand at a pivotal moment in history, like a grand symphony where various players must harmonize to create real progress. Generative AI is shaped by a diverse cast of stakeholders, each with distinct values, ambitions, and priorities, all striving to leave their mark on this emerging landscape. Each part matters, but none can carry the performance alone. True advancement will come when these groups align, so we can create a future that benefits all of us.

At the forefront are academics – the originators, the composers. They write the score, pushing the boundaries of what's possible. Every major advancement in AI has sprung from their efforts. But AI scientists are more than inventors – they are visionaries. They imagine not only what the technology *can* do, but what it *should* do. To disregard their vision is to undermine the very moral and intellectual foundation of the AI revolution.

Then there's industry – the performers. Corporations, particularly the major tech players and the startups they fund, take research and translate it into tangible products. They scale, shape, and bring these innovations to the public. But they also improvise, embellish, and, at times, distort the original score. They are always watching academia, scouting the next big idea to commercialize. This dynamic is not a flaw; it's how the system operates – but it underscores the power imbalance, as those who invent AI are often not the ones who determine how it's used.

Behind the scenes are investors – the conductors, guiding the tempo of innovation. With trillions in their control, they choose which ideas get the resources to grow and which are left in the dust. While they carry fiduciary responsibility, they also hold the power to greenlight or silence entire categories of innovation. And that power, while often exercised quietly, shapes the future. Startups respond accordingly – adapting their pitch, their product, and even their purpose to match the rhythm and priorities set by investors.

Finally, there's the public – the audience. Often seen as passive, they are far from it. Their responses shape market forces and drive regulation. They may not hold a baton or an instrument, but the entire performance is for them. People want technology that improves their lives – tools that empower, not overtake. They want to feel that the rapid march of progress is headed toward something worthwhile.

When these parts sync, the result can be extraordinary. When they don't, we face dissonance: AI rushed to market without oversight, and tools built for short-term profit rather than long-term flourishing. A discordant orchestra harms the very people it was meant to serve.

But there's a way through. The vision I invite us to embrace is one where our creative machines – our remarkable, intelligent, delightful AI-powered systems – are *humble*. A humble AI doesn't try to outshine us; it amplifies us. It makes us more creative, more capable, more fully ourselves. When AI enhances our thinking and expands our imagination, it empowers us in ways that endure, making us sharper and more inspired.

That's when AI becomes indispensable. That's when people experience lasting value – and they return, again and again. Trust deepens, loyalty grows, and commercial success follows.

Is this the kind of future science fiction promised? Perhaps not. But science fiction writers didn't invent creative machines – my colleagues and I did. We brought them to life. And we did so with a vision: to build AI that serves, supports, and elevates humanity.

The baton has been raised, the notes are unfolding. The future is being written in real time. And if we get this right – if we listen, if we attune to one another – we can create a world where AI and humanity move forward in concert. Now, it's time to listen to one another, and set the tune for a future where the entire capitalist stack prospers while humanity flourishes.

Epilogue

Writing this book has been an extraordinary journey. Through this extended reflection on the age of AI, I was confronted by a key realization: it's all about humanity. Just as Harold Cohen, who first introduced me to creative machines, realized that he was the true creative force behind AARON, *I now see that AI has always been – and will always be – all about us.*

We brought AI into existence, and we continue to create it. And who do we make it for? For ourselves, of course. Yet somehow, we have come to see AI as "other" – a foreign presence to fear or to worship. AI quickly becomes something to compete against, something to lose to. Much like our history is marked by fear of those who are different, we are now directing the same fear-filled attitudes toward AI.

But the truth is, it isn't AI we fear, but each other. We urgently need to reframe our most pressing AI problems as human problems – conflicts of interest, failures of collaboration, and a lack of shared vision among stakeholders. And these human problems, profoundly challenging as they may be, offer the only productive path forward.

Human history reveals that we've always struggled to build societies that are fair and sustainable. We have a habit of using innovation to benefit the few while making things worse for the many. But we also have a long track record of kindness, of helping each other, of learning and growing into our capacity to do good. That's where our greatest hope resides.

We must move beyond the two-camp approach – one camp claiming that AI is all glory, and the other, all folly. These perspectives are equally narrow and view this technological innovation as disembodied from the humanity that created it. As with any other major discovery, the opportunities for good and evil abound with this new technology. We must navigate this new age with our eyes open, willing to see this invention in all its shades – and avoid the impulse to defend it at all costs, or to deny it the merit that it is due.

As Eastern traditions teach, balance is key. The familiar Taoist yin and yang symbol offers a guiding light, the black and white parts beautifully interlacing with each other. AI is a thing of wonder, but it is not the answer to human existence. It is dangerous, but powerless without humans directing it.

AI is but one thread in the fabric of humanity. Let us not become so enamored with it that we forget each other. What matters most – what truly gives life meaning, purpose, and a sense of happiness – are the things we've always needed: our loved ones, our communities, and the essential necessities of food, water, and shelter.

Yet let us not become so afraid of AI that we forgo the chance to grow. Amid the relentless noise of modern life – the pressure to do more, have more, and be more – perhaps the greatest gift we can give ourselves is the time to slow down and truly gaze into the mirror that AI now holds before us.

When we dare to look, AI reveals humanity's shadow with startling clarity. It reflects our racism, our sexism, and the many other "isms" we would rather keep hidden. But this reflection doesn't have to be a condemnation, it can be an invitation. Rather than turning away, or rushing to blame the technology – or those who created it – we are offered a rare chance: to face ourselves honestly, and to grow.

Lean in. Gaze into the images AI generates. Get close enough to recognize the biases woven through its outputs – because they are reflections of the biases woven through us. If we have the courage to meet these shadows with open eyes, perhaps we can finally gain the self-awareness needed to break free from the cycles of hate and harm that have plagued so much of our history.

There is yet another invitation, a chance to evolve with the aid of humble creative machines. A wide range of systems can be built, not to replace us, but to profoundly elevate human capability. They can make us – not the AI – more intelligent and more creative. They can open new worlds of creative possibility and expand the reach of our imagination.

AI doesn't have to make us subservient. Instead, its brilliance can be used to help cultivate a more capable humanity. We must remain the captains, steering the course and directing the future of our species. AI can make us *more* human, not less.

We need people who care deeply about our shared future to take part in shaping this critical chapter in our history. So I invite you to get involved. Apply your heart and mind to the challenging problems outlined in this book. Don't get discouraged by their complexity.

Humanity has solved many a riddle more complex than the ones we face today. And we can, and must, do it again. My greatest hope is that, even though we cannot escape all harm from AI, we can harness it to bring profound and lasting benefit to our world.

Attempts to predict the future are, I believe, futile. It is far better to take an active role in shaping the world we wish to live in. The clay is still wet – and what we mold today will outlive us all.

We praise the spark,
Yet flinch from flame.
All the while,
Our human hands,
Still shape the clay –
And through the smoke,
A hidden path,
Arises from the fray.

Acknowledgments

Writing a book is much like climbing a mountain. It's been an incredible journey, challenging at times, exhilarating, and deeply transformative. This work was made possible by the support and contributions of many.

First and foremost, my deepest gratitude goes to my wonderful editor, Sunnye Collins. Her guidance shaped this book in ways that cannot be overstated. From the early stages to the most tangled moments of revision, her advice was always spot-on – offered with a kindness and steadiness that helped me navigate the complex terrain of making this book what I had hoped it could be. It would not be what it is without her.

Beyond my editor, my brother Ronen Ackerman was the first to read the manuscript from start to finish. His questions challenged my thinking, and his insights helped sharpen the book considerably. His honest, incisive feedback led to many substantive changes.

To my husband, David Loker – the foundation of every success I've had over the past 18 years. Your unwavering belief in all my endeavors keeps me going. Thank you for the countless nights of patience as I revised endlessly, and for a particularly thoughtful and rigorous edit of the manuscript.

A sincere thank you to my friend and colleague James Morgan for your notes on the book, and our ongoing conversations about art, AI, and creativity that have helped shape my thinking and, by extension, this book.

Thank you to my wonderful colleague Alison Pease for your thoughtful feedback and keen insights. Your support was very helpful in refining and enriching how I present our research community, along with a critical historical moment in our field.

A heartfelt thank you to Rafael Pérez y Pérez and Roisin Loughran for your invaluable input on several key sections of this book. Your stories added depth and authenticity, helping to bring our community vividly to life.

To my PhD student, Juliana Shihadeh – thank you for your sharp eye and thoughtful edits to a few key passages, particularly in the chapter on Brilliance Bias. Thank you to Jui Banik for helping to track down sources in the early days of this journey.

To everyone in my research community – colleagues and friends – Geraint Wiggins, Dan Ventura, Rafael Pérez y Pérez, Anna Jordanous, Alison Pease, Amílcar Cardoso, Oliver Bown, Tony Veale, and so many more: thank you for welcoming me with open arms. I've found a home in computational creativity. What I've learned from you has been nothing short of a paradigm shift. My life wouldn't be the same without our community, and this book wouldn't exist without you.

To my co-founders, David Loker and Chris Cassion: so much of what I understand about creative machines has come from our journey at WaveAI. Thank you for embarking on this adventure with me. It means the world.

This book owes its very existence to James Minatel at Wiley. Thank you for reaching out at exactly the right moment, and for believing in this project from the start. This has been one of the most meaningful endeavors of my life, and it began with your initiative. And finally, a warm thank you to the rest of the exceptional team at Wiley, especially Satish Gowrishankar. Your support and guidance throughout this process helped bring this book to life.

About the Author

Maya Ackerman, PhD is an internationally recognized leader in generative AI, charting the future of human creativity in an AI-powered world. As CEO and Co-Founder of WaveAI, one of the first companies to specialize in generative AI, she has helped empower millions of creators across the globe.

An Associate Professor at Santa Clara University, Dr. Ackerman's body of work includes over 50 peer-reviewed publications in artificial intelligence. She holds a PhD in Computer Science from the University of Waterloo and completed postdoctoral fellowships at Caltech and UC San Diego. Beyond her academic and entrepreneurial achievements, she is also a singer, pianist, and songwriter.

Named a *Woman of Influence* by the *Silicon Valley Business Journal*, Dr. Ackerman's work has been featured in *NBC News*, *NPR*, *New Scientist*, and more. She has spoken at leading institutions including Microsoft, Oxford, Stanford, Google, and the United Nations.

In this book, she takes readers on a riveting journey through the past, present, and future of creative machines. Drawing from her rare vantage point – at the intersection of academia, business, and the arts – she explores not only what AI can create, but what it means for us when it does.

References

Ackerman, M., Morgan, J., and Cassion, C. (2018). Co-creative conceptual art. *Proceedings of the Ninth International Conference on Computational Creativity (ICCC)*: 1–8.

Ackerman, M. and Shihadeh, J. (2024). The collective mind: exploring our shared unconscious via AI. In: *Workshop on Theory of Mind in Human–AI Interaction at CHI 2024*.

AlDahoul, N., Rahwan, T., and Zaki, Y. (2025). AI-generated faces influence gender stereotypes and racial homogenization. *Scientific Reports* 15 (1): 14449.

Allport, G.W., Clark, K., and Pettigrew, T. (1954). *The Nature of Prejudice*. Reading, MA: Addison-Wesley.

Arce, J.M.R. and Winkelman, M.J. (2021). Psychedelics, sociality, and human evolution. *Frontiers in Psychology* 12: 729425.

Art Vancouver. (2023). "Can animals be considered artists?" *Art Vancouver*, 23 February. https://www.artvancouver.net/post/can-animals-be-consid ered-artists#:~:text=Some%20animals%20create%20art%20naturally, like%20short%20V%2Dshaped%20passages (accessed 17 March 2025).

Backlinko. (2025). "ChatGPT/OpenAI statistics: how many people use ChatGPT?" https://backlinko.com/chatgpt-stats?utm_source=chatgpt. com (accessed 13 April 2025).

Baker, S., Hamm, S., Porter, T., and Cu-Porter, D. (2015). *Cognitive Cooking with Chef Watson: Recipes for Innovation from IBM & the Institute of Culinary Education*. New York. Sourcebooks.

Barron, F. (1955). The disposition toward originality. *The Journal of Abnormal and Social Psychology* 51 (3): 478.

Bian, L., Leslie, S.J., and Cimpian, A. (2017). Gender stereotypes about intellectual ability emerge early and influence children's interests. *Science* 355 (6323): 389–391.

Bian, L., Leslie, S.J., and Cimpian, A. (2018). Evidence of bias against girls and women in contexts that emphasize intellectual ability. *American Psychologist* 73 (9): 1139.

Boden, M.A. (1990). *The Creative Mind: Myths and Mechanisms*. London: Weidenfeld and Nicolson.

Boesch, C. and Boesch, H. (1990). Tool use and tool making in wild chimpanzees. *Folia Primatologica* 54 (1–2): 86–99.

Bradshaw, T. and Heikkilä, M. (2025). "Microsoft's $13bn OpenAI tie-up cleared by UK competition regulator." *Financial Times*, 5 March. https://www.ft.com/content/8f7fcaaf-2ae0-4c2f-a016-6f51d2f83cba?utm_source=chatgpt.com (accessed 14 March 2025).

Cascio, C.J., Moore, D., and McGlone, F. (2019). Social touch and human development. *Developmental Cognitive Neuroscience* 35: 5–11.

Cassion, C., Ackerman, M., and Jordanous, A. (2021). The humble creative machine. *Intentional Conference of Computational Creativity (ICCC)*.

Cheatley, L., Ackerman, M., Pease, A., and Moncur, W. (2022). Musical creativity support tools for bereavement support. *Digital Creativity*, 33 (1): 1–17.

Colton, S. (2012). The painting fool: stories from building an automated painter. In: *Computers and Creativity*, 3–38. Berlin, Heidelberg: Springer.

Colton, S. and Ventura, D. (2014). You can't know my mind: a festival of computational creativity. *Intentional Conference of Computational Creativity (ICCC)*: 351–354.

Colton, S., Cook, M., Hepworth, R., and Pease, A. (2014). On acid drops and teardrops: observer issues in computational creativity. *Proceedings of the 7th AISB Symposium on Computing and Philosophy*: 1–8.

Devine, P.G. (1989). Stereotypes and prejudice: their automatic and controlled components. *Journal of Personality and Social Psychology* 56 (1): 5.

Dzieza, J. (2023). "AI is a lot of work." *The Verge*, 20 June. https://www.theverge.com/features/23764584/ai-artificial-intelligence-data-notation-labor-scale-surge-remotasks-openai-chatbots (accessed 27 April 2025).

Ellingsen, D.M., Leknes, S., Løseth, G., Wessberg, J., and Olausson, H. (2016). The neurobiology shaping affective touch: expectation, motivation, and meaning in the multisensory context. *Frontiers in Psychology* 6: 1986.

Ferenczi, A. (2024). "The art and science of Alexander Mordvintsev." Kate Vass Galerie. https://www.katevassgalerie.com/blog/the-art-and-science-of-alexander-mordvintsev (accessed 19 April 2025).

Field, H. (2025). "Softbank set to invest $40 billion in OpenAI at $260 billion valuation, sources say." https://www.cnbc.com/2025/02/07/softbank-set-to-invest-40-billion-in-openai-at-260-billion-valuation-sources-say.html (accessed 14 March 2025).

Fritscher, L. (2023). "Carl Jung's collective unconscious theory: what it suggests about the mind." https://www.verywellmind.com/what-is-the-collective-unconscious-2671571 (accessed 20 April 2025).

Garcia, C. (2016). "Harold Cohen and AARON—a 40-year collaboration." Computer History Museum, 23. https://computerhistory.org/blog/harold-cohen-and-aaron-a-40-year-collaboration (accessed 13 June 2025).

Garcia, C. (2019). "Algorithmic music – David Cope and EMI". https://computerhistory.org/blog/algorithmic-music-david-cope-and-emi/ (accessed 28 April 2025).

Ghosh, S. and Caliskan, A. (2023). 'Person' == light-skinned, Western man, and sexualization of women of color: stereotypes in stable diffusion. In: *Findings of the Association for Computational Linguistics: EMNLP 2023*: 6971–6985. Singapore: Association for Computational Linguistics.

Goldstein, R., Vainauskas, A., Ackerman, M., and Keller, R. (2019). Brain-controlled musical improvisation. *International Conference on Computational Creativity (ICCC)*: 282–285.

Greenwald, A.G. and Banaji, M.R. (1995). Implicit social cognition: attitudes, self-esteem, and stereotypes. *Psychological Review* 102 (1): 4.

Greenwald, A.G., McGhee, D.E., and Schwartz, J.L. (1998). Measuring individual differences in implicit cognition: the implicit association test. *Journal of Personality and Social Psychology* 74 (6): 1464.

Hardus, M.E., Lameira, A.R., Van Schaik, C.P., and Wich, S.A. (2009). Tool use in wild orangutans modifies sound production: a functionally deceptive innovation? *Proceedings of the Royal Society B: Biological Sciences* 276 (1673): 3689–3694.

Harvey Mudd College. (2020). *Bob Keller | In memoriam*. https://www.hmc.edu/in-memoriam/bob-keller/ (accessed 28 April 2025).

Hu, K. (2023). "ChatGPT sets record for fastest-growing user base – analyst note." *Reuters*. https://www.reuters.com/technology/chatgpt-sets-record-fastest-growing-user-base-analyst-note-2023-02-01/ (accessed 14 March 2025).

Hunt, E. (2019). "Tay, Microsoft's AI chatbot, gets a crash course in racism from Twitter." *The Guardian*, 9 September. https://www.theguardian.com/technology/2016/mar/24/tay-microsofts-ai-chatbot-gets-a-crash-course-in-racism-from-twitter (accessed 28 April 2025).

i24NEWS. (2024). "UK, Canada record highest ever incidents of antisemitism in 2024." *i24NEWS*, 8 August. https://www.i24news.tv/en/news/international/diaspora-affairs/artc-uk-canada-record-highest-ever-incidents-of-antisemitism-in-2024 (accessed 21 April 2025).

Jandial, R. (2024). *This Is Why You Dream: What Your Sleeping Brain Reveals About Your Waking Life*. London: Penguin.

Johnson, G. (1997). "Undiscovered Bach? No, a computer wrote it." *The New York Times*. https://www.nytimes.com/1997/11/11/science/undiscovered -bach-no-a-computer-wrote-it.html (accessed 28 April 2025).

Jonker, F.A., Jonker, C., Scheltens, P., and Scherder, E.J. (2015). The role of the orbitofrontal cortex in cognition and behavior. *Reviews in the Neurosciences* 26 (1): 1–11.

Kaplan, J., McCandlish, S., Henighan, T., et al. (2020). Scaling laws for neural language models. *arXiv preprint arXiv:2001.08361*.

Kemmet, T. (2023). *The Magic of Mushrooms: Turning "Public Enemy Number One" Into an Ally to Help Put an End to the War on Drugs*. https://uclaw review.org/2023/02/14/the-magic-of-mushrooms (accessed 26 April 2025).

Kerner, S.M. (2024). "Stability AI gets new leadership as gen AI innovations continue to roll out." *VentureBeat*, 21 June. https://venturebeat.com/ai/ stability-ai-gets-new-leadership-as-gen-ai-innovations-continue-to-roll-out (accessed 28 April 2025).

Kraft, A. (2016). "Microsoft shuts down AI chatbot, Tay, after it turned into a Nazi". https://www.cbsnews.com/news/microsoft-shuts-down-ai-chatbot-after-it-turned-into-racist-nazi/ (accessed 29 April 2025).

Lambert, N. (2003). *A Critical Examination of Computer Art: Its History and Application*. Doctoral dissertation, University of Oxford.

Leslie, S.J., Cimpian, A., Meyer, M., and Freeland, E. (2015). Expectations of brilliance underlie gender distributions across academic disciplines. *Science* 347 (6219): 262–265.

Levine, P.A. (1997). *Waking the Tiger, Healing Trauma: The Innate Capacity to Transform Overwhelming Experiences*. Berkeley, CA: North Atlantic Books.

Livingston, P. (2005). *Art and Intention: A Philosophical Study*. Oxford: Clarendon Press.

Lopez, G. (2016). "Nixon official: real reason for the drug war was to criminalize black people and hippies." *Vox*, 23 March. https://www.vox.com/ 2016/3/22/11278760/war-on-drugs-racism-nixon (accessed 26 April 2025).

Loughran, R. and O'Neill, M. (2016). The popular critic: evolving melodies with popularity driven fitness. Musical Metacreation (MuMe), Paris. https://musicalmetacreation.org/mume2016/proceedings/Loughran_ the_popular.pdf (accessed 19 May 2025).

Love, A.W. (2007). Progress in understanding grief, complicated grief, and caring for the bereaved. *Contemporary Nurse* 27(1): 73–83.

Love, S. (2024). 'Like looking at a blank TV screen': what it's like to live with aphantasia. *The Guardian*. https://www.theguardian.com/wellness/2024/ feb/26/what-is-aphantasia-like (accessed 26 April 2025).

Lowe, R. and Leike, J. (2022). Aligning language models to follow instructions. https://openai.com/index/instruction-following/ OpenAI Blog (accessed 14 March 2025).

Marchese, D. (2018). "Seth Rogen on comedy, drugs, and creativity: an interview." *Vulture*. https://www.vulture.com/2018/04/seth-rogen-in-conversation.html (accessed 26 April 2025).

Marr, B. (2017). "Grammy-Nominee Alex Da Kid creates hit record using machine learning." https://www.forbes.com/sites/bernardmarr/2017/01/30/grammy-nominee-alex-da-kid-creates-hit-record-using-machine-learning (accessed 29 April 2025).

McIntyre, H. (2024). *PAUL McCartney reveals the Beatles' 'Yesterday' came to him in a dream*. https://www.forbes.com/sites/hughmcintyre/2024/02/22/paul-mccartney-reveals-the-beatles-yesterday-came-to-him-in-a-dream (accessed 16 March 2025).

Mok, A. (2022). "Google's management has reportedly issued a 'code red' amid the rising popularity of the ChatGPT AI." https://www.businessinsider.com/google-management-issues-code-red-over-chatgpt-report-2022-12 (accessed 14 March 2025).

Muradoglu, M., Arnold, S.H., Leslie, S.J., and Cimpian, A. (2023). "What does it take to succeed here?": the belief that success requires brilliance is an obstacle to diversity. *Current Directions in Psychological Science* 32 (5): 379–386.

Okasha, S. (2003). "Biological altruism." *Stanford Encyclopedia of Philosophy*. https://plato.stanford.edu/eNtRIeS/altruism-biological (accessed 1 June 2025).

OpenAI. (2022). "ChatGPT: optimizing language models for dialogue." https://openai.com/blog/chatgpt (accessed 27 April 2025).

Osorio, E.K. and Hyde, E. (2021). "The Rise of Anxiety and Depression among Young Adults in the United States - Ballard Brief." https://ballardbrief.byu.edu/issue-briefs/the-rise-of-anxiety-and-depression-among-young-adults-in-the-united-states (accessed 27 April 2025).

Oyama, S. (1976). A sensitive period for the acquisition of a nonnative phonological system. *Journal of Psycholinguistic Research* 5: 261–283.

Pearce, M. and Wiggins, G. (2001). Towards a framework for the evaluation of machine compositions. *Proceedings of the AISB* 1: 22–32.

Pease, A. and Colton, S. (2011). On impact and evaluation in computational creativity: a discussion of the Turing test and an alternative proposal. *Proceedings of the AISB symposium on AI and Philosophy* 39. York: Society for the Study of Artificial Intelligence and Simulation of Behaviour.

Peralta, E. (2013). "Researchers find that dolphins call each other by 'name.'" *NPR*, 21 February. https://www.npr.org/sections/thetwo-way/2013/02/

20/172538036/researchers-find-that-dolphins-call-each-other-by-name (accessed 17 March 2025).

Pérez y Pérez, R. (2017). *MEXICA: 20 Years—20 Stories*. Denver, CO: Counterpath.

Pérez y Pérez, R. and Sharples, M. (2001). MEXICA: a computer model of a cognitive account of creative writing. *Journal of Experimental & Theoretical Artificial Intelligence* 13 (2): 119–139.

Pérez y Pérez, R. and Sharples, M. (2023). *An Introduction to Narrative Generators: How Computers Create Works of Fiction*. Oxford: Oxford University Press.

Rees, A. (2004). "Nobel Prize genius Crick was high on LSD when he discovered the secret of life." *Mail on Sunday*, 8 August: 44–45.

Rogen, S. (2021). *Yearbook*. Crown: New York.

Rolls, E.T. and Grabenhorst, F. (2008). The orbitofrontal cortex and beyond: from affect to decision-making. *Progress in Neurobiology* 86 (3): 216–244.

Rubin, R. (2023). *The Creative Act: A Way of Being*. New York: Penguin Press.

SAG-AFTRA. (2024). "SAG-AFTRA strikes video games over A.I". https://www.sagaftra.org/sag-aftra-strikes-video-games-over-ai (accessed 27 April 2025).

Salinas, A., Shah, P., Huang, Y., McCormack, R., and Morstatter, F. (2023). The unequal opportunities of large language models: examining demographic biases in job recommendations by ChatGPT and LLaMA. *Proceedings of the 3rd ACM Conference on Equity and Access in Algorithms, Mechanisms, and Optimization*: 1–15. New York: Association for Computing Machinery.

Schwartz, E.H. (2023). "Riffusion AI Visualizes and Plays Music from Text Prompts – Voicebot.ai." https://voicebot.ai/2023/01/03/riffusion-ai-visualizes-music-from-text-prompts (accessed 19 April 2025).

Seth, A. (2017). Your brain hallucinates your conscious reality. *TED*. https://www.youtube.com/watch?v=lyu7v7nWzfo (accessed 27 April 2025).

Seth, A. (2022). "The big idea: do we all experience the world in the same way?" *The Guardian*, 8 November. https://www.theguardian.com/books/2022/oct/03/the-big-idea-do-we-all-experience-the-world-in-the-same-way (accessed 27 April 2025).

Sharples, M. (1996). An account of writing as creative design. In: *The Science of Writing* (eds. C.M. Levy and S. Ransdell), 127–148. London: Routledge.

Sharples, M. (2002). *How we Write: Writing as Creative Design*. London: Routledge.

Sharples, M. and Pérez y Pérez, R. (2022). *Story Machines: How Computers Have Become Creative Writers*. London: Routledge.

Shihadeh, J., Ackerman, M., Troske, A., Lawson, N., and Gonzalez, E. (2022). Brilliance bias in GPT-3. In: *2022 IEEE Global Humanitarian Technology Conference (GHTC)* (62–69). IEEE.

Shihadeh, J. and Ackerman, M. (2023). What does genius look like? An analysis of brilliance bias in text-to-image models. *Proceedings of the International Conference on Computational Creativity (ICCC)*.

Shihadeh, J. and Ackerman, M. (2024). Female professors grow beards: gender nonconformity in a Midjourney occupational analysis. *Proceedings of the International Conference on Computational Creativity (ICCC)*.

Singh, D., Ackerman, M., and Pérez y Pérez, R. (2017). A ballad of the Mexicas: automated lyrical narrative writing. *Proceedings of the International Conference on Computational Creativity (ICCC)*.

Smith, M. (2023a). "Burnout is on the rise worldwide—and Gen Z, young millennials and women are the most stressed." https://www.cnbc.com/2023/03/14/burnout-is-on-the-rise-gen-z-millennials-and-women-are-the-most-stressed.html (accessed 27 April 2025).

Smith, T. (2023b). "Stability AI fundraising leak: what we know." *Sifted*. https://sifted.eu/articles/stability-ai-fundraise-leak (accessed 28 April 2025).

Stanford HAI. (2021). *Rooting out Anti-Muslim bias in popular language model GPT-3.* https://hai.stanford.edu/news/rooting-out-anti-muslim-bias-popular-language-model-gpt-3 (accessed 20 April 2025).

Stein, M.I. (1953). Creativity and culture. *The Journal of Psychology* 36 (2): 311–322.

Sun, L., Wei, M., Sun, Y., Suh, Y.J., Shen, L., and Yang, S. (2024). Smiling women pitching down: auditing representational and presentational gender biases in image-generative AI. *Journal of Computer-Mediated Communication* 29 (1): zmad045.

Sundararajan, L. (2014). Mind, machine, and creativity: an artist's perspective. *The Journal of Creative Behavior* 48 (2): 136–151.

Sundararajan, L. (2021). Harold Cohen and AARON: collaborations in the last six years (2010–2016) of a creative life. *Leonardo* 54 (4): 412–417.

Tassi, P. (2024). "OpenAI removes AI voice that sounds like Scarlett Johansson in 'Her'." *Forbes*, 20 May. https://www.forbes.com/sites/paultassi/2024/05/20/openai-removes-ai-voice-that-sounds-like-scarlett-johansson-in-her (accessed 22 April 2025).

Turk, V. (2023). "How AI reduces the world to stereotypes." *Rest of World*, 6 February. https://restofworld.org/2023/ai-image-stereotypes (accessed 18 March 2025).

U.S. Department of Justice. (2023). "2022 FBI hate crimes statistics." https://www.justice.gov/archives/crs/highlights/2022-hate-crime-statistics (accessed 21 April 2025).

Vallor, S. (2024). *The AI Mirror: How to Reclaim Our Humanity in an age of Machine Thinking*. Oxford: Oxford University Press.

Valyaeva, A. (2023) "AI image statistics for 2024: how much content was created by AI." *Everypixel Journal – Your Guide to the Entangled World of AI*, 14 August. https://journal.everypixel.com/ai-image-statistics (accessed 20 April 2025).

Vaswani, A., Shazeer, N., Parmar, N., et al. (2017). Attention is all you need. *Advances in Neural Information Processing Systems* 30: 5999–6009.

Vincent, J. (2016). Twitter taught Microsoft's AI chatbot to be a racist asshole in less than a day. *The Verge*. https://www.theverge.com/2016/3/24/1129 7050/tay-microsoft-chatbot-racist (accessed 28 April 2025).

Vincent, J. (2019). "Microsoft invests $1 billion in OpenAI to pursue holy grail of artificial intelligence." *The Verge*, 22 July. https://www.theverge. com/2019/7/22/20703578/microsoft-openai-investment-partnership-1-billion-azure-artificial-general-intelligence-agi (accessed 14 March 2025).

Volz, K.G. and von Cramon, D.Y. (2009). How the orbitofrontal cortex contributes to decision making—a view from neuroscience. *Progress in Brain Research* 174: 61–71.

Wakefield, J. (2016). "Microsoft chatbot is taught to swear on Twitter". https:// www.bbc.com/news/technology-35890188 (accessed 28 April 2025).

Wiggins, G.A. (2006). A preliminary framework for description, analysis and comparison of creative systems. *Knowledge-based Systems* 19 (7): 449–458.

Witters, B.D. (2023). "U.S. depression rates reach new highs." *Gallup.com*, 26 March. https://news.gallup.com/poll/505745/depression-rates-reach-new-highs.aspx (accessed 27 April 2025).

Wohlleben, P. (2016). *The Hidden Life of Trees: What They Feel, How They Communicate—Discoveries from a Secret World*. London: Greystone Books.

Yong, E. (2022). *An Immense World: How Animal Senses Reveal the Hidden Realms Around us*. London: Random House.

Zweig, C. and Abrams, J. (eds.) (1991). *Meeting the Shadow: The Hidden Power of the Dark Side of Human Nature*. London: Penguin.

Index